THE CATTLEYAS AND THEIR RELATIVES

Volume III. *Schomburgkia, Sophronitis,* and Other South American Genera

Schomburgkia undulata. "Discovery of a plant!" From *Illustration Horticole* (1878).

THE CATTLEYAS AND THEIR RELATIVES

Volume III. *Schomburgkia, Sophronitis,* and Other South American Genera

A book in six parts
by

CARL L. WITHNER

TIMBER PRESS
Portland, Oregon

ISBN 0-88192-269-2
Printed in Hong Kong

TIMBER PRESS, INC.
9999 S.W. Wilshire, Suite 124
Portland, Oregon 97225

Library of Congress Cataloging-in-Publication Data
(Revised for vol. 3)

Withner, Carl L. (Carl Leslie)
The Cattleyas and their relatives.

Includes bibliographical references.
Includes indexes.
Contents: v. 1. The Cattleyas — v. 2. The Laelias — v. 3. Schomburgkia, Sophronitis, and other South American genera.
1. Orchids. 2. Cattleyas. 3. Orchid culture.
4. Orchids—Latin America. I. Title.
SB409.W793 1988 635.9'3415 88-8560
ISBN 0-88192-099-1 (v. 1)
ISBN 0-88192-161-0 (v. 2)
ISBN 0-88192-269-2 (v. 3)

Table of Contents

This volume is dedicated to all the
libraries and herbaria of the world
and the
devotion of generations that has made them
priceless depositories
of our present knowledge.

Preface

The material for Volume III has been difficult to assemble, physically as well as mentally. I had thought the rock laelias in Volume II were tough to work through, but this group of species has been worse. There are many reasons for this, not least of which is that many are still unknown in cultivation and rarely seen even in the wild. For many species, the herbarium record is extremely limited. Some have been confused with others, and few recent comparative studies or collections have been made. Also, many genera contain miniature plants that have less-than-spectacular flowers. They are perhaps of more interest to botanists than to orchid growers, the miniature *Sophronitis* being, of course, the exception.

With increasing interest in habitat survival and active interest in the conservation of all species, avid orchid hobbyists wish to see and know more about the lesser known taxa. More growers are interested in miniatures than ever before, and many are prime candidates for these curiosities. A more directed interest in the less common species will also make hobbyists more aware of what they have in their surrounding natural environment. In Brazil, where many of the species in this book are to be found, the emphasis is still on the great cattleyas and laelias. Most orchid nursery staffs have not heard of many of the genera, let alone the species. I was sorry to be reminded of this fact during my visit to Brazil in 1989, which turned up little that was different from 25 years ago. How does one raise consciousness about the smaller species?

The general run of taxa in this volume is not to be found in ordinary orchid books. Because much of the detailed information is difficult to obtain other than from type descriptions and gleanings from herbarium labels, no modern accounts are to be found on many species—habitats, pollinators, flowering season, anything—and several have not been collected by anyone for sometimes more than 50 or even 100 years. To search them out again, one by one, and compile, with herbarium voucher specimens, a checklist of the orchids of Brazil, so that the material is readily available for scholarly study, would indeed be a worthy job. At present, there is no single institution in Brazil where such a facility is available and being used. The situation is amazing, since Brazil is, after all, one of the countries richest in orchids in the world! In *Orchidaceae Brasilienses* Pabst and Dungs (1975, 1977) estimated that the country contained 2300 species in 191 genera, which is surely not a final count. Ecuador, not far from Brazil on the west, is already estimated to have more than 3000 species in a much smaller area—only the size of Colorado in the United States by way of comparison.

With environmental destruction now a major concern of all nations, provisions for national herbaria and development of national floras have an urgency that cannot today be ignored. In any case, compiling the information contained in this volume has forced me to take a more strictly botanical-technical approach in providing the information. In many cases, little data of historical or anecdotal interest is available to enliven anatomical or taxonomic details. Sometimes the plants and flowers of a given species have never been illustrated or photographed before; and in one or two cases even type specimens were never designated or preserved for the taxon. By collating the material in this volume, one at least has something to shoot at should there be other interpretations of the specimens or new information forthcoming.

The absence of information in horticulturally oriented books has compelled me to turn to more professional sources. A good botanical research library as well as a fine herbarium are absolutely essential in compiling and writing a book such as this. I am constantly amazed and pleased, in this connection, with the facilities of the library at the New York Botanical Garden and with the superior efforts of Marie Long, their research librarian, in finding type references for my compilations. Having before retirement been a professor of biology at the City University of New York, I often enjoyed my academic connections to the Botanical Garden. I also taught in the Garden's horticultural program and was in charge of the greenhouse orchid collections for four years, so I came to appreciate the completeness of its library long ago. I am ever thankful that its tremendous assets are available not only to me but to the public as well through the efforts of its cooperative staff.

I must also express my profound appreciation for the facilities and staff of the orchid herbarium and library at the Royal Botanic Gardens in England with particular thanks to Dr. Phillip Cribb, who is in charge of the orchid collections there. My thanks also go to William T. Stearn, of the British Museum, who gave me helpful information about the work of Robert Brown. Volume III has, as will its successors, required consulting with foreign resources, for there is no other way to gain firsthand experience with many of these species; some are limited to a single type specimen in the Lindley Herbarium at Kew or some other herbarium. I am always gratified by the enthusiastic and helpful cosmopolitan group of professional orchidists who give little heed to political or geographical boundaries when it comes to the world of orchids. Individuals in pursuit of orchid knowledge can thereby be citizens of the world. Our arguments can be limited to the discussions of points of anatomy or classification rather than politics—any fuss over boundaries is in regard to generic or species limits.

The same accolades must be extended to the facilities of the Orchid Herbarium of Oakes Ames at Harvard University and to Dr. Leslie Garay, now retired from the curatorship. The Ames has been a favorite place to visit and study over the years, and it has also been critical for compiling this volume, as it will be for those to come. The richness of the herbarium in New World material and its equally fine library of reference books are not to be duplicated, and Dr. Garay's uncanny knowledge of these plants is not to be matched. My appreciation also goes to Dr. Gustavo Romero, the present orchid specialist in charge of the Ames, who has been generous with his help and expertise in answering my many questions.

Evelyn Darrow is in Interlibrary Loan at the Wilson Library of Western Washington University in Bellingham, where I am now a research associate in the Biology Department. She continues to assist in my search for reference material concerning various species, and I thank her kindly. I would also extend my appreciation to the herbarium facilities at the U.S. National Herbarium at the Smithsonian Institution in Washington, D.C., and for the use of the herbarium at the New York Botanical Garden. Charlotte Tancin, librarian at the Hunt Institute for Botanical Documentation at Carnegie Mellon University, has helped me with early references in the Americas.

In addition, I have found the American Orchid Society judging records to be of help on many occasions when rare species have been singled out by judges for botanical or horticultural recognition or their owners cited for growing specimen-quality plants displayed at shows. The judging records do much to give an understanding of certain species, though as an American Orchid Society judge I hasten to add that this is not a plea to recognize horticulturally every orchid species as it comes along! We certainly do not need to worry that plants, so far collected only along a river on the top of Auyantepui in Venezuela, or found only once or twice on a mountainside in Brazil, will ever be a problem for the judges at future orchid displays!

On the other hand, orchid growers do have a distinct and varied role in conservation: the growing, showing, describing, reproduction, and distribution of unusual or rare species to other living collections, as well as the occasional preparation of herbarium material (especially after shows). Some hobbyists are already well aware of these responsibilities; others are not. We are not even certain what species are in cultivation around the world, what gems from the wild are being watered, fertilized, and otherwise looked after today. Our best clue is found in the listings of the species used in hybrid registration with the Royal Horticultural Society. According to some recent statistics, fewer than 2000 species of the estimated 30,000 in the orchid family have been used as parents for hybrids. Many, many more are present in our living herbaria, our collections—enormous assets when considered in their entirety!

A major problem for this volume has been finding illustrative material for all of the species discussed. Since many have never been illustrated in photographs or color plates or even grown in cultivation before, this difficulty is not surprising. It is, however, frustrating, since part of my responsibility is to find slides or pictures, no matter how elusive, to accompany the descriptions. Not only have I written many letters to helpful correspondents in various countries searching to find suitable photos, but files of the American Orchid Society have also been opened to me. I do appreciate the generosity of the many slide donors whose handiwork appears in this volume. All are credited at the pictures they provided, and Trudi Marsh must once again be singled out for special acknowledgment. When slides or plates could not be found, I have resorted to available drawings, all of which I have credited in the captions. Illustrations without a credit line are my own.

I must also thank my publisher, Timber Press, for the realization that orchid people are picture-minded, as Richard Abel, the former editor-in-chief, once said to me while in a mood to reveal state secrets. I am grateful that he has been willing to illustrate these volumes so thoroughly, making them so much more useful to all readers.

Artist Joy Dabney, now retired from the Media Department at Western Washington University, has prepared the *Schomburgkia* drawings from my sketches and from the dried flowers in my reference collection. These drawings make it possible for the first time to compare all these species at a glance. Joy has also prepared most of the biogeographical diagrams from my rough sketches, and I marvel at her interpretive abilities.

I am especially privileged to have received the permission from Greville Mee, the husband of the late Margaret Mee, to use her painting of *Sophronitis*. I was struck with this idea when I first saw a colored plate of the picture on sale at the Botanic Garden in Rio de Janeiro in January 1989. It is especially appropriate, since this particular water color was not included in the fascinating and beautiful book chronicling Margaret Mee's travels and artistry in Brazil, *In Search of Flowers of the Amazon Forests* (1988).

I must explain that it has been necessary to devote more time to deciding just which genera and which species should be included in this and subsequent volumes. I started this series with the intention of dealing with *The Cattleyas and Their Relatives,* in other words the whole of the Laeliinae. But the definition of Laeliinae, just what it includes and how far the idea should be taken in these volumes, soon arises. The major genera of many species present no problem. The difficulties arise in deciding which of the monotypic and minor genera should be included, especially if they have never yet been in cultivation. I have decided to err on the side of including too much rather than too little, both for the sake of completeness and also to raise awareness about many of these forms. Orchid growers have come to understand that neither plants nor flowers need be large to be fascinating, worth observing, and grown in our collections—and the smaller species fit better into small greenhouses. This particular volume concentrates on genera found mostly in South America, especially Brazil; later volumes will include Central American, Mexican, and Caribbean species of other minor genera.

Schomburgkia, as a genus, was last described in a 1913 monograph by Schlechter. *Sophronitis* endured a checkered career until the writings of Fowlie in recent years, but the genus has not been described as a group since the writing of Cogniaux in 1898–1902. Clearly, it is high time that these genera again be described, summarizing all that has been learned about the plants since those earlier days and mentioning the problems that are still unresolved. Since many of the difficulties can be remedied only by acquisition of new and better specimen material and observation of living plants in flower, field work continues to be critical. But, as I stated in the preface to Volume II, we may have no more than 10–20 years to do the bulk of the collecting before many habitats are effectively destroyed. We can be but pleased that many orchid species grow in mountainous terrain that is ill-suited for many practical purposes and is of limited accessibility.

The only modern book presenting a detailed study of the Laeliinae is the third edition of Schlechter's *Die Orchideen* (*Schlechter III*) by Brieger, Maatsch, and Senghas. Although much has been published in recent years about the overall classification of the Orchidaceae into families and subfamilies and tribes or subtribes, little or no attention has been given to the taxonomic categories below the subtribe level. The classification of the Laeliinae (Epidendrinae), a subtribe, must be worked out in some degree to make the choices necessary for the species included in this book. Brieger, once a student of Schlechter, spent many years teaching about and studying orchids in Brazil, and *Schlechter III* might be considered, in its way, to be a distillation of his ideas about Brazilian species. His coauthors, Maatsch and Senghas, have also had long experience with orchids through both academic and horticultural connections.

Although some are quick to point out deficiencies in the new *Schlechter III,* its treatment of and keys to the Epidendrinae are well worth consideration and study. The only current alternate system is provided by Dressler (1981), whose work is also outlined below. The two systems vary much in their detail and nomenclature though both are concerned with the same plants. Without compelling the reader to wade through all the technicalities of priority and validity of publication according to the International Code, we shall accept the systems as they have been published. The term Epidendrinae is used in the *Schlechter III* for this subtribe, but the alternate term, Laeliinae, is given priority as the subtribe epithet in Dressler's system. But whichever term is used, we can still communicate about the subtribe.

Brieger and his associates first divide the subtribe Epidendrinae (Laeliinae) into what could be called series, distinguishing four such series in the subtribe. Dressler, however, divides the subtribe Laeliinae (Epidendrinae) into six alliances (series?) instead of four. The two subtribal classifications are itemized below.

The four series and their main characteristics in the *Schlechter III* system presented by Brieger and his coauthors are as follows:

1. **Series Ponerae.** Column with an elongated foot; the lip free at its attachment to the column at the end of the foot; the lateral sepals free on either side of the lip and forming a prominent chin or nectary pouch with the lip base at the rear of the flower. This series includes the genera *Domingoa, Hexadesmia, Huebneria, Isochilus, Jacquiniella, Leaoa, Octadesmia, Orleanesia, Platyglottis, Ponera, Pseudohexadesmia, Pseudoponera, Scaphyglottis,* and *Sessilibulbum.*

2. **Series Hexiseae.** Column with an elongated foot to which the base of the lip is fused for a distance to produce a nectary pouch; lateral sepals also fused at their bases to the column and lip to help form the sides of

the pouch. This series includes the genera *Alamania, Amblostoma, Arpophyllum, Costaricaea, Diothonaea, Hexisea, Homalopetalum, Isabelia, Loefgrenianthus, Minicolumna, Nageliella, Neolauchea, Pachystele, Pinelia, Pygmaeorchis, Reichenbachanthus,* and *Stenoglossum.*

3. **Series Epidendrae.** Column without an elongated foot; the lip lying directly in front of column and parallel to it at its basal end; lip fused to the whole column, producing a nectary tube that becomes part of the fruit structure but seldom showing as a spur. This series includes the genera *Auliza, Epidanthus, Epidendrum, Nanodes, Oerstedella, Physinga,* and *Pleuranthium.*

4. **Series Cattleyae.** Column without an elongated foot; lip completely free of the column or only united with it for a short distance at the base. This series includes the genera *Artorima, Basiphyllaea, Barkeria, Broughtonia, Brassavola, Cattleya, Cattleyopsis, Caularthron* (*Diacrium*), *Constantia, Dimeranda, Dinema, Dressleriella, Encyclia, Hagsatera, Hormidium, Lanium, Leptotes, Microepidendrum, Neocogniauxia, Pseudolaelia, Renata, Rhyncholaelia, Schomburgkia, Sophronitella, Sophronitis,* and *Tetramicra.*

Although many recognize and approve of most of these genera, others object to details of this grouping by Brieger, Maatsch, and Senghas. But, as I noted above, theirs is the only published modern account with keys and diagrams to guide the researcher, and I find the system helpful. For those interested in a complete collection of the Epidendrinae (Laeliinae), their classification provides an awesome if not fearsome checklist with which to compare the completeness of collections. A total of 60 genera could be represented, only a few of which are monotypic.

None of the species of the Brieger, Maatsch, and Senghas series of Epidendrae will be included in these volumes. First of all, no one person presently knows enough to be authoritative about all the Epidendrae and everything else, too. *Epidendrum,* the major genus in series 3, for example, has dozens of species, many yet unnamed. Second, I am limited to six volumes! The series Epidendrae will demand its own specialist and books at some future time. And, one could say, my time is running out! The coverage that I provide will be a sampling from the balance of the subtribe, and I only hope that I do not leave out someone's favorite species or the next curiosity that takes the fancy of the judges or collectors.

The Dressler system has been presented as part of his overall classification scheme for the orchid family. He divides the subtribe Laeliinae (Epidendrinae) into six alliances and 43 genera. Until now, however, the tentative alliances have only been outlined, not named, nor does Dressler provide any keys for separating them or the genera. No doubt this information will appear in the future, and then we will have a definitive alternative to the *Schlechter III* system detailed above. Meanwhile, for comparison with the above system, I include Dressler's six alliances and the genera each contains:

Alliance 1. *Helleriella, Hexisea, Isochilus, Nidema, Platyglottis, Ponera, Reichenbachanthus,* and *Scaphyglottis.*

Alliance 2. *Artorima, Basiphyllaea, Brassavola, Broughtonia, Cattleya, Constantia, Encyclia, Hagsatera, Isabelia, Laelia, Pseudolaelia, Quisqueya, Rhyncholaelia, Schomburgkia, Sophronitis,* and *Tetramicra.*

Alliance 3. *Barkeria, Caularthron,* and *Orleanesia.*

Alliance 4. *Dimeranda, Diothonaea, Epidanthus, Epidendrum, Jacquiniella, Neowilliamsiana,* and *Oerstedella.*

Alliance 5. *Alamania, Domingoa, Homalopetalum, Leptotes, Loefgrenianthus, Nageliella,* and *Pinelia.*

Alliance 6. *Dilomilis* and *Neocogniauxia.*

One readily observes the differences and similarities of the two systems, so we shall await the amplification of Dressler's system with interest. Using either system, the scope of these volumes in *The Cattleyas and Their Relatives* cannot possibly include all the genera, and I have finally had to make choices.

With this volume now completed, I look forward to writing the rest of the volumes and the research and study that must go into them. I have always liked this sort of puzzle solving. It is one of the things I do best in science, and I hope that all orchidists will understand this subtribe to a greater degree as a result of the research and writing. Also, I know now why I have been taking so many pictures and begging so many flower samples over the years. Once again, my appreciation goes out to all those individuals who have helped, directly and indirectly.

March 1992

CHAPTER 1

The Genus *Schomburgkia*

HISTORY OF THE GENUS AND ITS CULTURE

Schomburgkia seems to have been a confusing genus since it was first established by Lindley in 1838 in *Sertum Orchidaceum, A Wreath of the Most Beautiful Orchidaceous Flowers*. The genus included two species at that time, *S. crispa* and *S. marginata*. Both are handsomely pictured in Lindley's book, and *S. crispa* was the type species for the genus. To quote from Veitch's *Manual* (1887):

> The genus is dedicated to Dr. Richard Schomburgk, a distinguished botanist, who accompanied by his brother, Sir Robert Schomburgk, as naturalist in the boundary expedition to British Guiana, in 1840–44, during which he discovered the typical species, *S. crispa,* and numerous other orchids until then unknown to science. Dr. Schomburgk has been for many years Director of the Botanic Garden, at Adelaide, South Australia.

In 1840 Lindley described another species of *Schomburgkia, S. superbiens,* but identified it as a *Laelia*. This error has produced, down through the years, a whole series of scientific papers by various orchid taxonomists on whether the two genera should be kept separate or combined. Let us hope that the publication of *this* monograph may, at least, lay that vintage problem to rest. Lindley went on to name other species in this genus as *Schomburgkia,* apparently never knowing what controversy he had initiated.

Another confusion dates from 1917 when Rolfe set up the new genus *Myrmecophila* for those species of the genus with hollow pseudobulbs that harbor ants. The ants, by virtue of their ferocious nature and sharp mandibles, apparently protect the plants from would-be robbers of nectar or pollen, gaining a living space in return. The choice of using the segregate genus for the ant-inhabited taxa, or keeping them all together within

Schomburgkia, seems to me a relatively arbitrary one. After many discussions, the debate would seem to come down to the old argument of the splitters vs. the lumpers. True, the *Myrmecophila* are mostly lowland plants, and they do have the specialized "cow horn" pseudobulbs and lack large floral bracts, but the flowers and growth habits are essentially similar to those of their more upland, solid pseudobulbed cousins. I view these differences at the subgeneric level, not the generic one, so will have to be counted among the lumpers on this occasion. My answer to all of these difficulties is to use the epithet *Schomburgkia* to describe all of this genus and deal with the above differences at the subgeneric level. For those who may wish to use the epithet *Myrmecophila,* I have starred the proper name listed among the binomial synonyms of the various species. There are no proper *Myrmecophila* names for *Schomburgkia albopurpurea* or *S. sawyeri* yet published, so that anyone wishing to use that genus for these species will have to publish the transfer from *Schomburgkia.*

A group of taxa with easily recognized characteristics, the schomburgkias are large and sun-loving (photophilic); have tough, hard, usually ridged pseudobulbs of several, more or less equally spaced nodes (homoblastic) that are covered with closely fitting bracts; two, three, or four leathery, firm, stiff leaves; elongated flower stalks with close, papery bracts and terminal clusters of flowers.

The sepals and petals usually have undulate or crisped margins and, except for the lip, are similar in shape and appearance. It should be noted, however, that in most of the species the sepals tend to be wider and longer than do the petals, as in some bifoliate cattleyas. The column is often exposed, lying against the lip, with the lip prominently decorated with wavy keeled, or at least raised, veins that obviously serve to help attract and guide pollinators. In some cases the column matches the color of the edges of the lateral lobes and sometimes contrasts with the lateral lobes, which enclose it to varying degrees. There are eight pollinia—two equal sets of four—in the anther.

The taxa of this genus have an interesting distribution pattern with most of the fusiform-pseudobulbed species concentrated in the northern portions of South America, Venezuela, and Colombia. The hollow-pseudobulbed species are, with one exception in South America, all found in Middle America or the Caribbean. Overall, the genus stretches from the southern Andes (Bolivia and Peru) north into Central America and México, and from there into the islands of the Caribbean. Plants of most species are found at sea level or at relatively low elevations not exceeding 1500 m.

Plants of all the species grow in the most exposed situations, often along riversides on rock, or above swampy areas on old tree snags, where sunlight, heat and humidity are paramount. Blooming them in cultivation not only requires a large space for the plants with their elongated flower stalks, but one must also provide the requisite heat and light.

Schomburgkias are very "deliberate" plants that hate to be repotted or disturbed and may just sit for two or three years doing "nothing" after transplanting. Once they are established and growing, one tries to keep them going by not changing their growing conditions, especially if they begin to flower on a regular basis. These plants are ideal for naturalizing in warm climates on various trees around a garden or home. If left undisturbed, they can form large clumps with many long roots running up and down a tree trunk, and they will flower every year.

They are best adapted to greenhouse culture by cultivation in baskets or on a horizontal slab of treefern, suspended near the glass where light is maximal and air circulation is best. Let them grow two or three seasons over the side of the container for best flowering. If pot grown and the plant-over-the-edge bothers you, just slip pot and all into one or two larger pots without disturbing anything in any way. As the roots grow they will fill the spaces between the pots and hold them tightly together.

A good measure of *Schomburgkia* culture, besides the size of the pseudobulb, is the color of pseudobulbs and leaves. Both should have a distinct yellow cast if sufficient light is being provided. The routine presence of flowers is, of course, the ultimate measure of successful culture. In general, the plants can grow well in a greenhouse with cattleyas or laelias under intermediate temperatures, the mountainous species requiring a little less heat and fresher air circulation than those found at sea level.

Little can actually be written about hybridization with this genus. Although a great variety of hybrids have been produced (see Table I for intergeneric combinations), most are of comparatively recent origin, and thus we know little about them. Mostly they are curiosities resulting from experimental crosses in an effort to produce as many new generic categories as possible. Most of the time the schomburgkias have been incidental in the process, not used specifically for any characteristics they might possess. It is intriguing that one of the larger species in this subtribe, *Schomburgkia crispa,* has been successfully combined with one of the smallest, *Sophronitis cernua.* It should also be noted that one of the intergeneric combinations with *Schomburgkia* is of six different genera—*Johnyeeara.*

Four species have been used more than others in the hybrids: *Schomburgkia lyonsii, S. thomsoniana* (var. *minor*), *S. superbiens,* and *S. undulata.* While the habit of the plants is dwarfed or reduced by the combination with other parents, the schomburgkia qualities come through strongly. The flowers, though reduced in number, are usually clustered at the top of an elongated flower stalk, and they retain their ruffled, random appearance. The colors also come through strongly or lend an intensity to the new hybrids. Also the photophilic nature of the plants is soon noticeable, and most are readily grown with ample light and warmth.

Cleistogamy has been observed in certain species of this genus—the production of seed pods before the flowers even open and without benefit of insects. This process is initiated by the premature withering of the rostellum so that the stigmatic fluids may reach the pollinia without the pollen being formally transferred during pollination. This has been particularly observed in *Schomburgkia lyonsii* and *S. wendlandii,* and Jones (1967a) has observed it also in *S. humboldtii, S. marginata, S. undulata,* and *S. crispa.*

There are more species of *Schomburgkia* than most people suspect. Accordingly, they may be mislabeled and confused in cultivation. Check your plants carefully against the descriptions and illustrations here to help put the correct names on them. A key is provided in Table I and outline drawings of flower parts in Figure 1-1. Relative size is the easiest way to help differentiate among the species after noting the color of the flowers, pedicellate ovaries, bracts, columns, and the anther caps. Next would come the number and position of the keeled veins and the degree of lobing of the lip. As with other orchids, knowing the country of origin is of special value in identification as it may immediately eliminate many species from consideration. You will note that many species have only a localized distribution (the Pacific coast of México, Venezuela, Cayman Brac Island, and so on), but unfortunately you may have no way of knowing this until after the plant is identified. Readers of Volumes I and II of this series will realize the importance I have placed on geography in determining species limits within this subtribe of orchids. Although a given genus may be of ancient evolutionary origin, the multiplicity of taxa within it can point to more recent speciation by introgression, polyploidy, or other mechanisms. Endemism is often the rule with orchid species, and apparently for some species the distribution is very limited. Whether this is the result of comparatively recent origin, limited survival from ancient times, or the result of geographic or other barriers cannot at this point be determined beyond

TABLE I

Intergeneric Combinations with *Schomburgkia* Registered Through September, 1992, with Approved Abbreviations

SCHOMBURGKIA	
× *Brassavola*	= *Schombavola (Smbv.)*
× *Brassavola* × *Broughtonia* × *Cattleya* × *Laelia*	= *Westara (Wsta.)*
× *Brassavola* × *Cattleya*	= *Dekensara (Dek.)*
× *Brassavola* × *Cattleya* × *Laelia*	= *Recchara (Recc.)*
× *Brassavola* × *Cattleya* × *Laelia* × *Epidendrum*	= *Yahiroara (Yhra.)*
× *Brassavola* × *Cattleya* × *Laelia* × *Epidendrum* × *Sophronitis*	= *Johnyeeara (Jya.)*
× *Brassavola* × *Cattleya* × *Laelia* × *Sophronitis*	= *Fergusonara (Ferg.)*
× *Brassavola* × *Laelia*	= *Maclemoreara (Mclmra.)*
× *Broughtonia*	= *Schombonia (Smbna.)*
× *Broughtonia* × *Cattleya*	= *Schombocatonia (Smbcna.)*
× *Broughtonia* × *Cattleya* × *Epidendrum*	= *Wilburchangara (Wbchg.)*
× *Broughtonia* × *Diacrium*	= *Shipmanara (Shipm.)*

conjecture. But here such isolated populations are mostly treated as separate species. By my present count the genus *Schomburgkia* is composed of 22 species.

Schomburgkia is absolutely one of the most perplexing orchid groups with which I have worked to date, and had I not been collecting their flowers at every show for the past 25 years for my little index card herbarium and taking photos of the plants as well, I doubt that this account could be as accurate as I believe it is. There is little or no good herbarium material for many of the species—almost nothing, literally, to study or consult. Often type descriptions are lacking in critical details as well. It is with some trepidation that I have written the accounts on the following pages, but with the good help of those at the New York Botanical Garden Research Library to track down type descriptions and other pertinent literature, and the willing assistance provided by others in the form of slides or specimens, more data on this group has been assembled than before. I hope that the species descriptions will call forth still more information for a better knowledge of this fascinating yet difficult group. I feel certain that it will one day be found to hold a central role in the evolution of the Laeliinae.

× *Broughtonia* × *Laeliopsis*	= *Hildaara (Hdra.)*
× *Cattleya*	= *Schombocattleya (Smbc.)*
× *Cattleya* × *Diacrium*	= *Mizutara (Miz.)*
× *Cattleya* × *Diacrium* × *Laelia*	= *Mailamaiara (Mai.)*
× *Cattleya* × *Epidendrum*	= *Scullyara (Scu.)*
× *Cattleya* × *Epidendrum* × *Laelia*	= *Northenara (Nrna.)*
× *Cattleya* × *Epidendrum* × *Laelia* × *Sophronitis*	= *Izumiara (Izma.)*
× *Cattleya* × *Laelia*	= *Lyonara (Lyon.)* 1959 not 1948
× *Cattleya* × *Laelia* × *Sophronitis*	= *Herbertara (Hbtr.)*
× *Diacrium*	= *Schombodiacrium (Smbdcm.)*
× *Epidendrum*	= *Schomboepidendrum (Smbep.)*
× *Epidendrum* × *Laelia*	= *Dillonara (Dill.)*
× *Laelia*	= *Schombolaelia (Smbl.)*
× *Laelia* × *Sophronitis*	= *Deiselara (Dsla.)*
× *Sophronitis*	= *Schombonitis (Smbts.)*

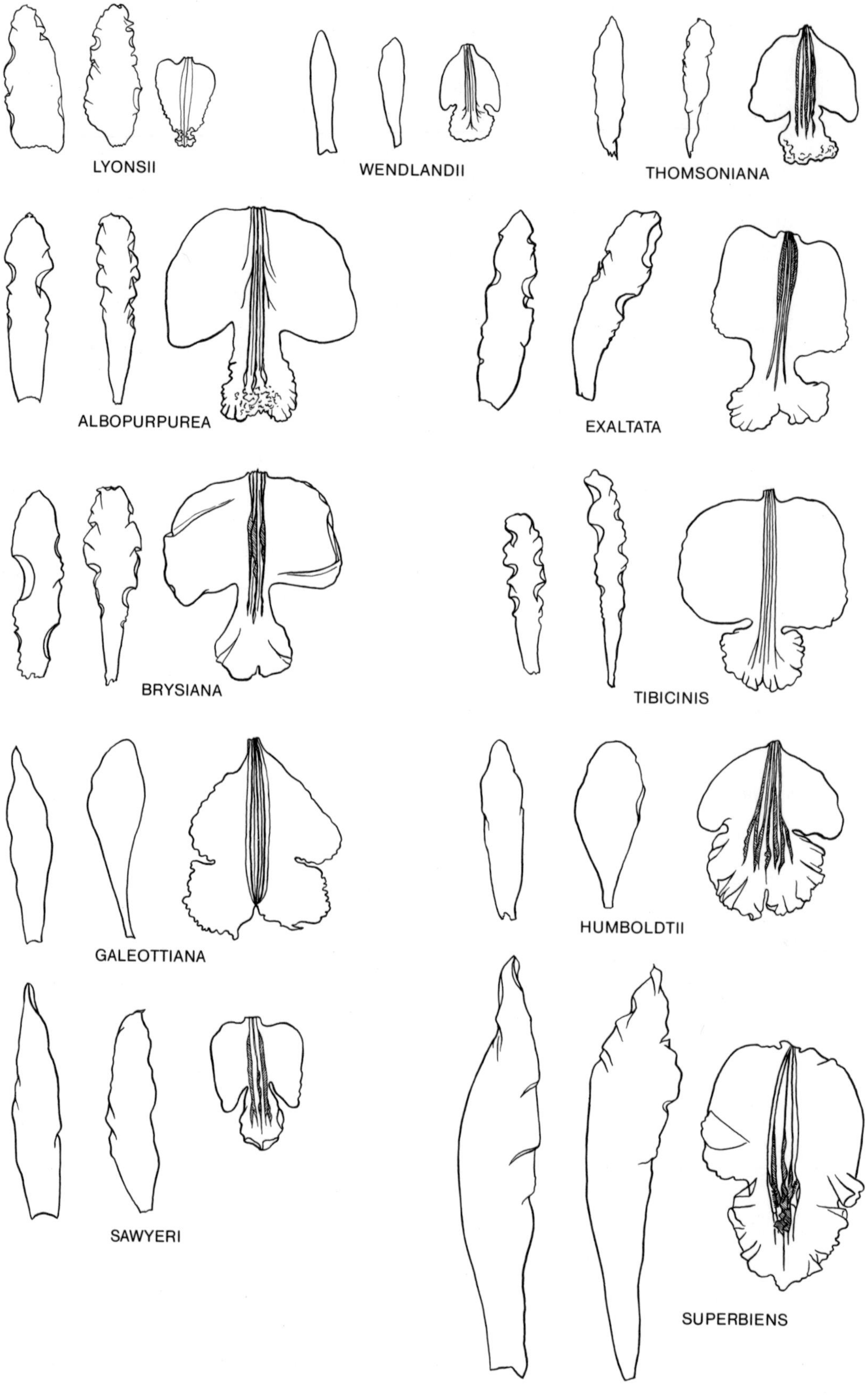
LYONSII
WENDLANDII
THOMSONIANA
ALBOPURPUREA
EXALTATA
BRYSIANA
TIBICINIS
GALEOTTIANA
HUMBOLDTII
SAWYERI
SUPERBIENS

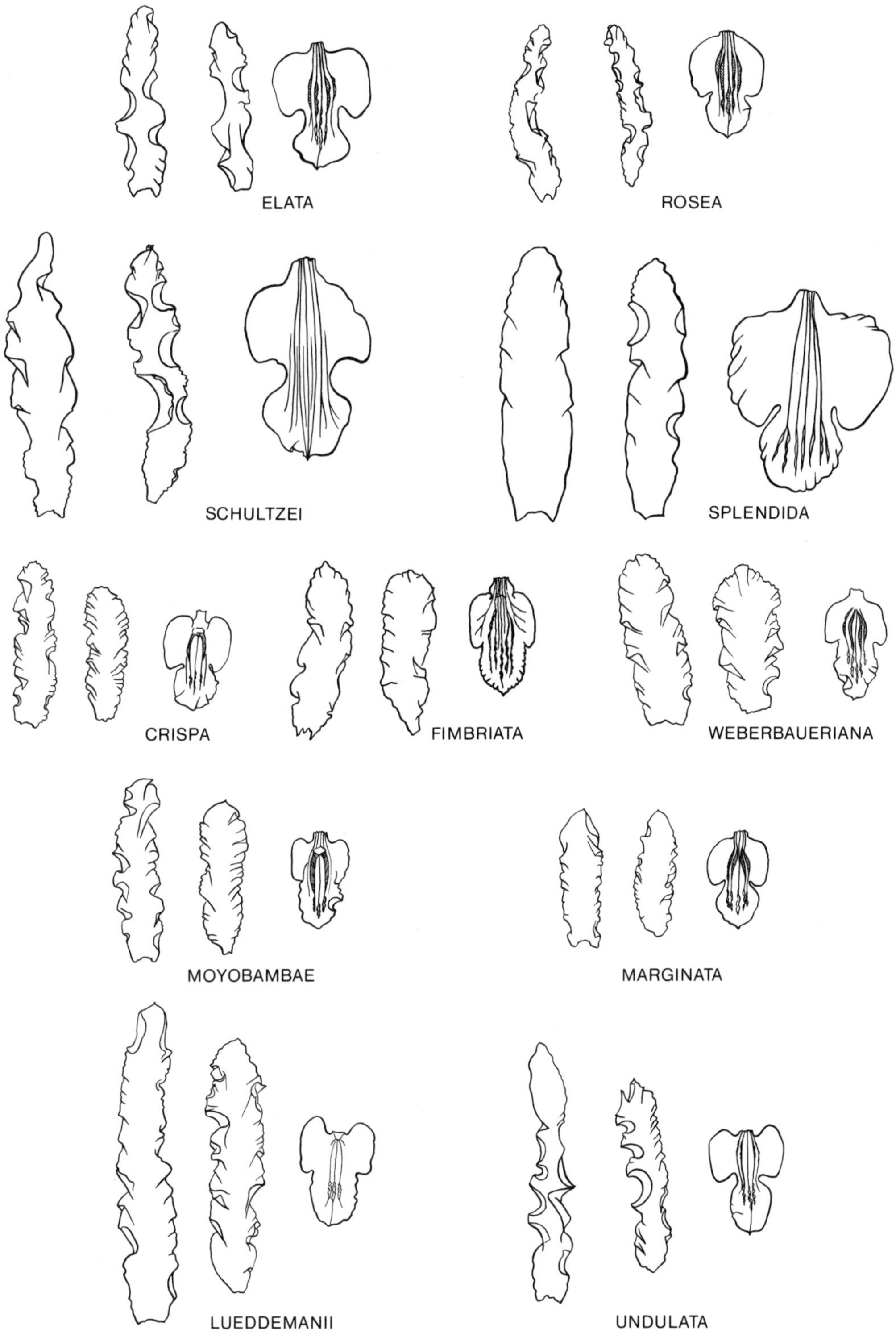

Fig. 1-1. Outline drawings of dorsal sepal, petal, and lip of *Schomburgkia* species. All sizes relative. Drawings made by Joy Dabney from actual flowers.

CLASSIFICATION OF THE SPECIES

Schomburgkia may readily be subdivided into two and possibly three or four subgenera based on pseudobulb habit, flower color, and then lip anatomy. Two of the species have non-resupinate flowers, an unusual situation in an otherwise generally consistent taxon. Some orchidists have suggested that this feature could be the basis for a third subgenus, but since one of the non-resupinate species, *S. lyonsii,* has solid fusiform pseudobulbs, and the other, *S. wendlandii,* has hollow "cow horn" pseudobulbs, it is not logical to establish a subgenus that cuts across the characters of the other two subgenera. One could consider the pseudobulb habit as more primitive or primary in the evolution of the genus than the flower position in these two taxa. Pseudobulb habit involves all of the species, while the flower position involves only two species. The latter may be viewed as a localized response to specific pollinators present in a given geographical area. Limiting the third subgenus to only one species, *S. lyonsii,* is also not useful at present.

Schlechter in 1913 was the first to monograph the genus and provide a résumé of all the species then known, saying that they were clearly separated from the genus *Laelia*. He also pointed out that Reichenbach had lumped them with laelias and other plants into the genus *Bletia* in 1861 but later changed his mind and properly recognized their original classification as *Schomburgkia*. Schlechter recognized 13 species and two hybrids, dividing the species into two sections, but missed *Laelia superbiens* (1840) and *L. wendlandii* (1861) completely in his inventory.

His first section, *Eu-Schomburgkia,* was characterized by comparatively short flower stems with thick, almost umbel-like clumps of flowers terminating a long stalk and long, prominent floral bracts. The *Eu-*, meaning the true schomburgkias, is not necessary according to present rules of nomenclature, making it *Schomburgkia* today; and, instead of making it a section, we make a subgenus that includes the type species for the genus, *S. crispa*.

Schlechter's second section was *Chauno-Schomburgkia* with sometimes branched flower stalks and a successive production of flowers with short bracts. Today we know it as *Chaunoschomburgkia*. *Chauno-* implies a gap or cleft and refers to the hollow pseudobulbs of this subgenus, each developing an opening at its base through which ants may pass in and out. Rolfe in 1917 proposed generic status for this taxon, calling it *Myrmecophila,* the ant-lover. Although this genus epithet has been used from time to time by some authors, it is not usually accepted, so that the included species are all mostly still called *Schomburgkia*. *Chaunoschomburgkia* has the epithet priority for this subgeneric category of the genus. As noted earlier, some individuals prefer to use the genus *Myrmecophila*. They could ignore this discussion of subgenera, since no subdivision would be necessary.

Although Schlechter, in the years after his monograph, went on to describe four more species of *Schomburgkia* from 1913 to 1924, he never published another account of the whole genus equivalent to his 1913 paper. Further, when he published a similar paper on the genus *Laelia* in 1917, he included the section *Calolaelia* (beautiful laelia) with the *L. superbiens* of Lindley without recognizing that it should have been in his 1913 article on *Schomburgkia. Laelia wendlandii* was included in the 1917 paper as a species of uncertain placement—possibly as a species of the section *Podolaelia*—so he also missed it as a species of *Schomburgkia,* though it had been known since 1861.

There is nothing like hindsight to solve certain taxonomic problems. Although many of the difficulties with this genus seem obvious points to us now, previous orchidologists may have had only one cultivated plant or a single herbarium specimen to study. Often the records upon which today's opinions are based do not consist of much more than the identical specimen studied earlier. But usually we have a few firsthand field observations, a photo or two, and frequently some cultivated plant accounts, perhaps even judging records, to amplify the available information. As tropical environments are destroyed in many localities, the records available become all the more critical. We need to support efforts for enhancing the continuation of species populations, if not in the wild then in cultivation. You will find, as the species are described in coming pages, that the records on many schomburgkias are very incomplete and need more and better information. Do not hesitate to send in any observation you have made, especially if it differs from what you read here.

At this point a new classification system is called for—or at least a more detailed division of the genus beyond dividing it into two subgenera—*Schomburgkia* and *Chaunoschomburgkia.* I feel, however, that the state of our knowledge is so deficient within the group that it cannot yet be effectively carried out. Accordingly, I will describe certain affinities that are apparent and let it go at that. Some of these tentative generic subgroups or sections have only a single species.

Schomburgkia lyonsii is unique in the remarkably reduced size of its midlobe which, along with its non-resupinate and spotted flowers, must have to do with coadaptation with pollinators. This midlobe feature is not found in other taxa of the genus, though often the lateral lobes of the lip in various species are considerably larger than the midlobe. This species is the only one with fusiform pseudobulbs to be found native in the Caribbean.

Schomburgkia wendlandii is unique in its small, plain, non-resupinate flowers and short flower stalks. No other species is like it.

Schomburgkia humboldtii and *S. galeottiana* share a similar appearance with greatly expanded midlobes and almost unruffled cattleya-like petals wider than the sepals. Yet one species is Venezuelan and the other is Mexican—a disjunctive distribution. *Schomburgkia humboldtii* is the only South American species among the hollow-bulbed taxa.

The fluted species complex involves *Schomburgkia tibicinis, S. brysiana, S. exaltata, S. albopurpurea,* and *S. thomsoniana.* These all have large lateral lobes of the lip with the midlobe ranging from having no stalk at all (sessile) to having a distinct isthmus.

The "beautiful laelia" section, *Calolaelia* of Schlechter, contains *Schomburgkia superbiens* with its large flowers and intricate keels and lip configuration. It is the largest and most distinctive species in the genus. Its closest relative is *S. sawyeri,* which should probably be classified in the same section.

The *Schomburgkia crispa* complex includes yellowish brown or tan colored species that are difficult to distinguish clearly: *S. crispa, S. fimbriata (gloriosa), S. marginata, S. moyobambae,* and *S. weberbaueriana*. Geography will play a certain role in the separation of these entities, and their distributions in nature are not yet accurately known. The lips of this complex are uniformly pale and have several parallel keels. Sepals and petals tend to be shorter and broader than their counterparts in the following groups.

The *Schomburgkia lueddemanii* category has only one species: *S. lueddemanii*. The flowers are larger and more frilled than those of the above species but also yellow brown in general color. The lips are variously colored, particularly around the rim of the midlobe in contrast to the central portions and the disc. So far as can be stated today, this is the only schomburgkia species that is found in both South America and Central America, Venezuela to Costa Rica. This group may also include *S. wallisii,* listed below under questionable species.

The *Schomburgkia undulata* alliance includes *S. undulata, S. rosea, S. elata, S. schultzei,* and *S. splendida*. These species have dark brown-red-purple sepals and petals with magenta lips, usually fair- to large-sized flowers sometimes with a "varnished" appearance, and very curly segments. The floral bracts are also generally rose or pink color and thus add to the show. *Schomburgkia moyobambae* possibly belongs with this group—having the pink bracts and purple lip—instead of the *S. crispa* complex above.

The poorly known or questionable forms would include two species. The Venezuelan *Schomburgkia schlechterana* is still a largely unknown taxon. Although it belongs with the *S. crispa* or the *S. undulata* complex based on its lip structure, it is best now to consider it separately until more is known of its distribution and characteristics. Also poorly known because it lacks a critical type description and a type specimen or illustration is *S. wallisii* from Colombia. My best opinion at this writing would be that it was the Colombian form of *S. lueddemanii,* and it is therefore not included in the above groups. The epithet becomes a synonym and is really a *nomen nudum*.

These points of comparison will be discussed in more detail in the following pages. For further information, please refer carefully to the accompanying drawings of the lip outlines with their veining and keel patterns and to the silhouettes of the other flower parts (Fig. 1-1). Table II, a numbered key, may help in your identification of unlabeled plants.

TABLE II

Key to Species of *Schomburgkia*

1. Pseudobulbs solid, flowers in umbel-like terminal clusters, floral bracts nearly as long or longer than the pedicellate ovaries (Subgenus *Schomburgkia*) . . . . go to 2
1. Pseudobulbs hollow, flowers in succession along a sometimes branched stalk, floral bracts short (*Chaunoschomburgkia*) . . . go to 17
2. Flowers white, spotted purple, non-resupinate, sometimes incomplete . . . . *S. lyonsii*
2. Flowers otherwise . . . . go to 3
3. Dorsal sepal 80–90 mm long. *S. superbiens*
3. Dorsal sepal shorter . . . . go to 4
4. Flowers greenish or yellowish to tan-brown . . . . go to 5
4. Flowers red-brown, maroon or purple . . . . go to 11
5. Dorsal sepal 45–60 mm long and 5–7 mm wide . . . . go to 6
5. Dorsal sepal 35 mm or less in length. . . . . go to 7
6. Dorsal sepal to 60 mm, lip 3-keeled, pale with brown markings on the edges . . . . *S. lueddemanii*
6. Dorsal sepal 45–50 mm, lip 5-keeled on disc and pale pink overall. . . . . *S. fimbriata*
7. Pedicellate ovary 50 mm or less in length . . . . go to 8
7. Pedicellate ovary 65 mm long or longer . . . . go to 10
8. Lip white or pale, bracts papery white. . . . . go to 9
8. Lip and bracts rose pink . . . . *S. moyobambae*
9. Sepals and petals overall chestnut brown . . . . *S. crispa*
9. Sepals and petals edged with yellow . . . . *S. marginata*
10. Flowers green-gold color with dorsal sepal 8 mm wide. . . . . *S. schlechterana*
10. Flowers yellow brown color with dorsal sepal 10 mm wide . . . . *S. weberbaueriana*
11. Flowers small, dorsal sepal not over 25 mm long . . . . *S. rosea*
11. Flowers larger, dorsal sepal 30–40 mm or longer. . . . . . .go to 12
12. Dorsal sepal 10–15 mm wide. . . . . go to 13
12. Dorsal sepal not more than 9 mm wide . . . . go to 14
13. Yellow on disc, 5 equal white keels . . . . *S. splendida*
13. Lip and disc magenta, veins darker, raised but not keeled to tip of lip . . . . *S. schultzei*
14. Lip with central yellow area on disc; ovary and pedicel to 65 mm long . . . . *S. wallisii*
14. Lip without yellow spot, ovary and pedicel 70 mm or longer . . . . go to 15
15. Bracts not pink, becoming brown or papery . . . . *S. elata*
15. Pink floral bracts . . . . go to 16
16. Midlobe of lip with isthmus . . . . *S. undulata*
16. Midlobe of lip without isthmus. . . . . *S. heidii*
17. Flowers creamy white, clear or tawny yellow except for purple on lips and sometimes as a flush on backs of sepals . . . . . . go to 18
17. Flowers mostly purple though sometimes marked with tawny colors, occasionally a dull green flushed with maroon, sepals and petals not clear cream or yellow . . . . go to 20
18. Central American plants, lips with 5–7 keels, strongly undulate edges on sepals and petals, sepals curving forward at the tips . . . . *S. brysiana*
18. Cayman Island plants, lips 3-keeled, sepals and petals comparatively flat, at most wavy-edged. . . . . go to 19
19. Plants dwarf, flowers canary yellow with maroon-purple lip, found mostly on Cayman Brac Island, Cuba. . . . . *S. thomsoniana*

TABLE II *continued*

19.	Plants and flowers larger, creamy white with semi-open sepals and petals, deep purple lip, found on Grand Cayman Island	*S. albopurpurea*
20.	Flowers small, non-resupinate, dull green with white or pale lips	*S. wendlandii*
20.	Flowers medium or large, resupinate, magenta or brighter colored	go to 21
21.	Midlobe comparatively small, colors varied purple and tawny orange, petals narrower than dorsal sepal	*S. tibicinis*
21.	Midlobe expanded, large, ruffled, and rounded, sepals and petals all purple, petals wider than sepals	go to 22
22.	Venezuelan plants	*S. humboldtii*
22.	Central American-Mexican plants	go to 23
23.	Plants from Gulf-side Guatemala, Belize, upright flower stalks, sepals and petals pale on reverse	*S. exaltata*
23.	Plants from Pacific slope of Guatemala and México	go to 24
24.	Inflorescence nodding, 2-leaved plants, lip with 5–7 raised veins on midlobe	*S. sawyeri*
24.	Inflorescence upright, plants 3–4 leaved, lip with 3–5 raised veins on midlobe	*S. galeottiana*

DESCRIPTION OF SPECIES

List of species of *Schomburgkia*

S. albopurpurea
S. brysiana
S. crispa
S. elata
S. exaltata
S. fimbriata
S. galeottiana
S. humboldtii
S. lueddemanii
S. lyonsii
S. marginata
S. moyobambae
S. rosea
S. sawyeri
S. schultzei
S. splendida
S. superbiens
S. thomsoniana
S. tibicinis
S. undulata
S. weberbaueriana
S. wendlandii

Doubtful species

S. schlechterana
S. wallisii

White and Purple Schomburgkia

Grand Cayman Island

(PLATES 1, 2)

Schomburgkia albopurpurea (Strachan *ex* Fawcett) Withner. *Stat. nov.*
Basionym: *S. thomsoniana* var. *albopurpurea* Strachan *ex* Fawcett. 1894. *Bull. Bot. Dept., Jamaica,* n.s., 1:74.
Subgenus *Chaunoschomburgkia*

Synonyms

Schomburgkia thomsoniana var. *atropurpurea* Hooker f. 1902. *Curtis's Bot. Mag.,* t. 7815.
Schomburgkia brysiana var. *atropurpurea* (Hooker f.) Jones. 1963. *Am. Orch. Soc. Bull.* 32(1):4–12.

I have debated a while over how to handle this species concept since it has for years been mixed up so closely with another, *Schomburgkia thomsoniana*. After the latter species was named in 1887, it was subsequently divided in 1894 into two varieties, var. *minor* and var. *albopurpurea*. The *S. albopurpurea* was changed to var. *atropurpurea* when *S. thomsoniana* was written up by Hooker f. in 1902, likely by mistake or misinformation. To sort out this confusion I have decided to raise the two varieties to species status because I believe that doing so best represents the status quo of their biology. Please see the discussion under *S. thomsoniana* to read about the counterpart of this species.

These two species are native to the Cayman Islands, *Schomburgkia thomsoniana* proper coming from Little Cayman and Cayman Brac and *S. albopurpurea* coming from Grand Cayman. However, the *S. albopurpurea* forms on Grand Cayman have been naturally hybridized with *S. thomsoniana,* so that a mixed population of the two species results on that island, a fine hybrid swarm that should be carefully studied as an example of introgression in action. One should check their pollinators, time of flowering, specific habitats, details of flower structure, and colors of flower parts to understand how this process has occurred. The main islands are about 160 km apart, south of Cuba. The species *S. thomsoniana* has also been found on Cuba to which the Caymans were once attached in ancient geological times.

The flowers of *Schomburgkia albopurpurea* are creamy white, semi-open, and have a contrasting dark purple, almost black midlobe on the lip. The apex of the lip is usually contorted and recurved with many granulations and verrucosities over the veins of the whole surface of the tip. The plants have upright flower stalks with as many as 20 flowers produced in succession, only 3 or 4 open at any given time. Sometimes they appear as white tubes with a purple tip because of their semi-open habit (Plate 1). There is a good color illustration of an opened flower of this species in Kennedy's *Orchid Digest* article (1979), but it is inaccurately captioned *S. thompsoniana* (and with a *p*) (see Plate 2).

The plants are reported to grow on scrubby trees and are even exposed to the salt spray due to the proximity of the ocean on these small islands. The plants are, of course, fully photophilic and require a basket hung high in the greenhouse in order to grow properly and produce flowers in cultivation.

Jones has named the natural hybrid between *Schomburgkia atropurpurea* and *S. thomsoniana* as *S. brysiana* var. *intermedia*. The problem, as mentioned elsewhere, is that Jones repeatedly considered *S. brysiana* and *S. thomsoniana* to be the same species, whereas they are completely different, though both are yellow in color. The name for these hybrid forms from Grand Cayman Island would therefore properly be *S.* ×*intermedia* (Jones) Withner *comb. nov.*, and the concept of *S. brysiana* should not be considered at all in this connection. The basionym for this change would be *S. brysiana* var. *intermedia* (Jones, 1972c). The revised parentage of the hybrid plants would accordingly be *S. albopurpurea* × *S. thomsoniana*. A xanthotic clone with no red pigmentation that has been awarded by the American Orchid Society under the name *S. thomsoniana* 'Nono' is probably an example of this *S. intermedia*.

Some *Schomburgkia* hybrids in cultivation have been registered with one parent as *S. thomsoniana*. It is not likely that *S. albopurpurea* or its natural hybrids were ever confused in those registrations with *S. thomsoniana*. Since hybridizers have searched out the dwarfer, brighter, more open yellow forms with flatter, red-purple lips (*S. thomsoniana*) to use in their combinations, no parentage revisions are necessary as a result of raising variety *albopurpurea* to species status.

Brys's Schomburgkia

Belize, Costa Rica, Guatemala, Honduras, Nicaragua

(PLATES 3, 4)

Schomburgkia brysiana Lemaire. 1851. *Jard. Fleur* (Gand). 1:misc. 34, pp. 32–34.

Subgenus *Chaunoschomburgkia*

Synonyms

Schomburgkia campecheana (Lemaire) Kraenzlin. 1903. *Gard. Chron.*, 3rd ser., 34:381.

Schomburgkia brysiana var. *intermedia* Jones. 1972c. *Bol. Soc. Broteriana*. 46:244.

**Myrmecophila brysiana* (Lemaire) Kennedy. 1979. *Orchid Digest* 43(6):204–211.

Schomburgkia brysiana and *S. tibicinis* have a long history of being confused with one another. They do have similar appearing plants, and the flowers also have similar shapes, but there are distinctions when one knows what to look for. The flowers of *S. brysiana* have been termed the orange or yellow *S. tibicinis,* indicating one major differentiating characteristic—the color. Lemaire described the color of the flowers as a uniform ocher-orange in the type description. Another distinction is that the midlobe of the lip has a distinct claw or isthmus at its base in contrast to no claw in *S. tibicinis*. Also, there are three major raised or elevated veins on the disc of *S. brysiana* while *S. tibicinis* has five or seven, and in *S. brysiana* the small column is nearly enclosed by the lateral lobes of the lip.

These differences are spelled out and partly diagrammed by Garay and Slocum (1963). Horich (1988) has also written of the comparison between the two species in *Die Orchidee,* with color illustrations. Correll confused the two species as synonyms in *Orchids of Guatemala* (Ames and Correll, 1953), and those using this book must keep it in mind when reading the description of *Schomburgkia tibicinis*. That description, incidentally, also includes *S. exaltata* and identifies them all as species of *Laelia*. Jones, on a different tack, has completely confused this species with *S. thomsoniana* from the Caribbean and has thereby altered not only its characteristics but also its range of distribution. This particular error has been perpetuated by Brieger and his coauthors in *Schlechter III*. In the Rodriguez book, *Generos de Orquídeas de Costa Rica* (1986), this species is beautifully illustrated as *S. tibicinis* with a note on the plate by the artist that it is called *guarion amarillo*. There is apparently no debate about the synonymy of *S. campecheana* with this species, it having been likened to *S. thomsoniana* in early articles. Enough said about this particular can of worms!

There may be two to four leaves on the pseudobulbs, and the flower stalk may or not be branched to 1.5 m in length. The 9–32 blossoms are produced in succession. Lemaire says that the flowers are larger than the usual *Schomburgkia tibicinis* and equivalent in size to *S. tibicinis* var. *grandiflora* but others write that plants and flowers are smaller. The ocher-orange ground color of the flower parts of my specimen shows flushes of purple along the sepal and petal margins and especially at the petal tips. The central keeled veins are yellow, and the lateral veining in the tube of the lip is purple. The crisped and indented midlobe of the lip can be flushed with darker purple. The ovaries and column are white, sometimes with a flush of pink or violet.

The geographic range of this species is still open to question because of the confusion with *Schomburgkia tibicinis,* but it would seem to be on the Caribbean side of Central America. The species occurs in Guatemala and Honduras and, based on the Rodriguez plate, in Costa Rica. One assumes it may also be found in Nicaragua or even into Panama. The plant I have flowers in the United States in summer and was collected by Chris and Helen Turnbull in Guatemala where it flowered in March. There is no mention of this species in Hamilton's data because it is mostly not described or recognized, or it is possibly confused with *S. thomsoniana.*

The original *Schomburgkia brysiana* plant and three other orchids were sent by Arnold Brys, a merchant from Bornhem, near Anvers, France, for identification to Charles Lemaire, a professor of horticulture and an orchidologist as well. The schomburgkia was a new entity, but we are not told how Brys obtained the plant or where it came from. The spray was 162 cm long and bore 32 flowers!

Jones (1963) has named a presumed natural hybrid, *Schomburgkia parkinsoniana,* based on a single plant, with the assumed parentage being *S. humboldtii* × *S. brysiana.* He does not explain the crossing of a Central American plant with one from Venezuela.

Jones (1967b) has also described *Schomburgkia rechingerana* as a possible hybrid between *S. tibicinis* × *S. brysiana,* both Central American-Mexican species, based, again, on a single plant found on Barbados in the West Indies with no clues as to its origins. His description cites a flower that is pink outside, yellow-brown within, and has a lip of dull cream with a white forelobe and five-keeled veining. Kennedy (1979) also mentions natural hybrids between these two species and illustrates one in his *Orchid Digest* article, though it does not match Jones's description. The concept may, nonetheless, be a valid one. There may indeed be intermediates between these two Central American species, but one could question how they got to Barbados and their assignation there. Since another plant with larger flowers than ordinary was named *S. rechingerana* var. *superba* by Jones (1968b), there must have been at least two plants in this hybrid population on Barbados.

Wavy or Crisped Schomburgkia

Colombia, Guyana, Suriname, Venezuela

(PLATES 5, 6)

Schomburgkia crispa Lindley. 1838. *Sert. Orch.*, Plate 10 and *Bot. Reg.* 30, t. 23.

Subgenus *Schomburgkia*

Synonyms

Schomburgkia marginata Hooker (*non*-Lindley.). 1839. *Curtis's Bot. Mag.* 14, n.s., t. 3729.

Cattleya crispa Beer (*non*-Lindley.). 1854. *Prakt. Stud. Fam. Orch.*, p. 209. Not *L. crispa* Reichenbach f. nor *C. crispa* Lindley.

Bletia crispina Reichenbach f. 1861. *Walp. Ann. Bot.* 6:419.

Laelia gloriosa Williams. 1941. *Darwiniana* 5:76. Not *S. gloriosa* Reichenbach f.

Schomburgkia crispa is the type species for the genus, and it has one of the shortest Latin type descriptions of any orchid in these volumes—eight words: *petalis sepalis labellique lobo medio transverso obtuso crispis*. Lindley goes on to say, in English, that *S. crispa* has a stout, round rhizome with scars where the scalelike bracts have fallen off, fusiform very hard pseudobulbs 9–15 cm high with deep ridges covered closely by membranous sheaths. There are two tough, leathery leaves about 23 cm long × 5–6 cm wide with acute tips.

The scape is about 90–100 cm long covered completely with closely attached scales. Each flower is subtended by a spreading scaly bract; it is about 5 cm long but shorter than the ovary and pedicel. The flowers are in a dense cluster at the top of the stalk. These characteristics can be duplicated almost exactly, with some variation in sizes and colors, for most of these fusiform species in the subgenus *Schomburgkia*, so it is, indeed, a good type for the genus.

The flowers are about 5–6 cm in natural spread with stalk and ovary that together measure 5 cm in length. The sepals and petals are a light tan-yellow to brown-orange color with pronounced wavy and crisped margins. Their tips are blunt or rounded, and the sepals tend to be wider than the petals. (Readers may recall that the latter is also a characteristic of some of the bifoliate cattleyas described in Volume I of this series.) The lip is white to pale pink with the midlobe darkening toward the strongly reflexed tip. The sessile, transversely oval midlobe is about one-quarter to one-third the length of the lip and is transversely oval and much crisped, the lateral lobes obscurely set off from it. When the lip is flattened, it is nearly as wide as the distance across the lateral lobes. The type drawing shows that the column has wings tapering nearly to the base where the lip is attached.

There are five raised veins crested somewhat on the disc that are easily traced from the base of the lip. A raised central cushion or thickening is found toward the base of the lip and involves the central vein. The two smaller lateral veins on either side swerve to make up the margins of this area so that it fits within the winged edges of the column. This configuration would force the proboscis of the pollinator upward against the column in the quest for nectar at its base. The stalked ovary is white with green flushes, or it may have a pink flush on the upper portions as is shown in the type illustration. The floral bracts are soon dried, white, and chaffy.

According to the account in the original description, Schomburgk found the plants growing on the banks of the Corentyn River, and later the species was found to be more common along the Berbice River, both in Guyana. Subsequent reports would indicate that it is one of the most widespread of all the taxa in this genus, growing in Brazil, Ecuador, and Peru as well as in those countries listed above. However, this broad distribution may be the result of misidentifications, and only Colombia, Guyana, and Venezuela are certain at this time. The confused identifications are, perhaps, also indicated by the variety of flowering dates in Hamilton, from February to August, with concentrations in both February and August. Plants of the species may just bloom twice a year, but no one has ever mentioned that as an observation, and it would be a unique characteristic in the whole genus. A likely possibility is that some of the reports of this species are the closely related *Schomburgkia fimbriata (gloriosa)* that has often been confused with *S. crispa*. According to Dunsterville and Garay (1976), these two species should be kept as separate entities. Further discussion will be found under *S. fimbriata*. The new *Schlechter III* (Brieger et al., 1976–1981) lists *S. marginata, S. gloriosa, S. weberbaueriana,* and *S. moyobambae* as synonyms under *S. crispa,* an excessive condensation with which most orchidologists could not agree.

This species is a lowland riverine schomburgkia and likes heat, humidity, and sun. It first flowered in England in the Brocklehurst collection, near Macclesfield, from a specimen brought from Suriname by Lance.

Tall Schomburgkia

Colombia

(PLATES 7, 8)

Schomburgkia elata Schlechter. 1924. *Fedde Repert. Spec. Nov.* 27:138.
Subgenus *Schomburgkia*

Synonym

Schomburgkia imperialis. An unpublished herbarium name.

Schomburgkia elata is a species without a valid synonym, indicating in this genus that it was described at a later date and is practically unknown. The species is endemic in Colombia but is ordinarily not seen in our collections or even in theirs. Kennedy (1979) dismisses it with a single sentence, saying he collected it near Villavicencio, which is south of Bogotá on the Amazonian side of the Eastern Cordillera. He illustrates the flowers of the collected plant in his article. The type specimen was collected in gallery forests at 400 m near San Martín, Colombia, by Schultze (after whom *S. schultzei* is named). It was flowering in February 1920 when it was collected.

The sepals and petals are dark red-brown, somewhat broad but short, undulate, and curled back at their ends. The lip is a pale rose pink or magenta with 5–7 white keeled veins on the disc of the midlobe that is expanded and ruffled coming to the apex. The midlobe has a distinct isthmus, and there is no central white area as on the lips of *Schomburgkia undulata* or the yellow of *S. splendida*. The keels originate near the base of the lip, not just over the disc as in most species. The lateral lobes curl back on their edges and leave the pink column exposed. The pedicellate ovary is also pink, and the bracts are palely colored and soon become dried and brown.

Jones (1968d) has reduced this species to synonymy under *Schomburgkia undulata,* a questionable move, since the flowers obviously differ from that species in their size, coloration, keel pattern, and proportions. Even though we may need more material to study before giving secure definition to the concept of *S. elata,* the species is retained here as a separate taxon. In the past it has particularly been confused with *S. splendida,* but that species has larger flowers and a yellow disc on the midlobe. *Schomburgkia schultzei* does not have keels on the disc, only thickened veins.

The sepals are about 33 mm long in the type, the petals a little shorter. The lip is lobed at the middle, is generally oval-shaped when flattened, and is attached to the base of the column for about 4 mm. The lip measures 27 mm long and 20 mm wide across the lateral lobes. The midlobe is circular in outline and bluntly pointed, 12 mm long and wide. The curved column is narrowly winged and about 15 mm long, with the pedicellate ovaries measuring 80–85 mm. The name was chosen because of the extra tall 1.5 m flower stalks, a point of distinction for this species in the *S. undulata* complex.

Long-stalked Schomburgkia

Guatemala, Honduras, México

(PLATE 9)

Schomburgkia exaltata Kraenzlin. 1926. *Mitteil. Inst. Allg. Bot. Hamb.* 6(2):342.

Subgenus *Chaunoschomburgkia*

Synonyms

Schomburgkia tibicinis Bateman var. *exaltata* (Kraenzlin.) Jones. 1965. *Adansonia* 5(1):41–48.

**Myrmecophila exaltata* (Kraenzlin.) Kennedy. 1979. *Orchid Digest* 43(6):204–211.

The exalted name of this schomburgkia refers to the fact that the flowers are borne on high, not to some esoteric position in the genus or orchid kingdom hierarchy. It was named by Kraenzlin in 1926 after a plant collected by Ulmcke along the Rio Dulce near Livingston, Guatemala. This plant was collected on 11 April 1925 and sent to the Botanical Garden in Hamburg, Germany, where it flourished until the Garden was destroyed by bombing during WW II. The species has subsequently been collected in Chiapas, México, and has a distribution limited to northern Guatemala and adjacent southern México on the Caribbean side. According to older records, the flower stalks may be 2.25 m tall and bear up to 80 flowers.

In contrast to the flowers of *Schomburgkia tibicinis,* the flowers of *S. exaltata* have a lip with a tubular base, and the lateral lobes completely enclose the column. The mid- or apical lobe of the lip is broadly expanded, and there are three pronounced keeled veins extending down onto the white or creamy disc. Two other veins are not so prominent. The lateral lobes are purple tipped and purple veined over a white base, the keeled veins are yellow, and the midlobe is darkest purple. The sepals and petals have undulate edges with deep magenta-purple surfaces, sometimes with an orange overcast flush. The reverse sides are white flushed with a paler purple, and they contrast where the rippled edges approximate each other over their darker fronts. The columns are white.

The plants have thickly cylindrical, strongly ribbed, short pseudobulbs to 30 cm. They are three-leaved and can produce branched flower stalks that reach 2.25 m. As many as 80 flowers may be produced in succession. Pedicel and ovary measure 3.5–4 cm, sepals are 4 × 25 mm, petals are 35 mm long and 12 mm wide at their tips from 4 mm wide bases, and the lip is 25–27 mm long.

This species is placed in synonymy under *Schomburgkia tibicinis* by Ames and Correll in *Orchids of Guatemala* (1953), a procedure we would not follow today. Their description lumps this species together with *S. tibicinis* and *S. brysiana*. Jones continued the error, making this species a variety. In my opinion they are all three best considered as separate species. Confusion over the identity of *S. exaltata* has given rise to some of the reports of natural hybrids between *S. brysiana* and *S. tibicinis,* which could be part of the ultimate explanation for the origin of *S. exaltata* as a species. The riddle might be solved with more field investigation and better knowledge of specific distribution patterns. Some pollination experiments, both hybridizing and selfing of selected clones, might also provide useful information.

Glorious Schomburgkia

Brazil, Guyana, Suriname, Venezuela

(PLATES 10, 11)

Schomburgkia fimbriata (Vellozo) Hoehne. 1952. *Arq. Bot. Estado São Paulo* 2:142.

Subgenus *Schomburgkia*

Synonyms

Epidendrum fimbriatum Vellozo (*non*-H. B. K.). 1827. *Flora Flum.* 9:4.
Schomburgkia marginata Hooker. 1840. *Curtis's Bot. Mag.*, t. 3729.
Schomburgkia marginata var. *immarginata* Hooker. 1840. *Curtis's Bot. Mag.* 13, plate 3729.
Schomburgkia crispa Lindley. 1844. *Bot. Reg.* 30: t. 23 (*non*-Lindley, 1838, *Sert. Orch*).
Schomburgkia gloriosa Reichenbach f. 1860. *Hamb. Gartenz. Blumenzeit.* 16:178.
Bletia gloriosa (Reichenbach f.) Reichenbach f. 1861. *Walp. Ann. Bot.* 6:419.
Laelia gloriosa (Reichenbach f.) Williams. 1941. *Darwiniana* 5:76.
Schomburgkia amazonica Hoehne. *Nomen nudum*. Herbarium name.

Until I assembled and checked all the synonyms for the proper name of this schomburgkia, I had not realized that the basionym for the species was the Vellozo name of *Epidendrum fimbriatum*. The more common modern name, *Schomburgkia gloriosa* by Reichenbach, thus becomes a synonym, and anyone with a plant of this species will have to add a new label. *Schomburgkia gloriosa* was previously made a synonym by Schlechter in his monograph (1913) but under *S. crispa* and with a question mark. He was apparently not certain what to do with it or where to place it in his scheme. Hoehne made the proper combination of *S. fimbriata*, presumably not realizing that his *S. amazonica* was the same thing, but he did not have a chance to resolve this research before his death.

We must question the distribution outside of Venezuela and Brazil pending more field research on authentic specimens and more properly identified plants in cultivation for study. The flowers are usually confused with those of *Schomburgkia crispa* (please see) so that identification records must be double checked. *Schomburgkia fimbriata* has also sometimes been confused with *S. marginata*, but the yellow edges on the sepals and petals of the latter species are readily noted.

Quoting from Dunsterville and Garay (1976), "Reichenbach was correct in separating *Schomburgkia gloriosa* from *S. crispa* . . . the chief differences are in the shape of the lip and the column." They go on to describe these differences in detail and illustrate them with a fine drawing typical of that work. Pabst (1977) tells us that on checking all the Brazilian plants ascribed to *S. crispa*, he found they were in reality *S. gloriosa*. This would include the controversial *S. amazonica* Hoehne, an epithet that was never formally published.

The pseudobulbs are well spaced on the rhizomes, reaching a height of 30 cm and consisting of 4–6 nodes. The 2–3 rigid leaves are up to 6 × 32 cm with the flower stalk reaching 75 cm and bearing 8–15 flowers. The thin sepals and petals are brown or chestnut and can have a purple tinge with tesselated veins. The lip is pale with purple or brown flushes, especially toward the margins, and has amethyst veining in the throat. The lip is "very little three-lobed and broadest at the base" Reichenbach wrote. He went on to state that he was responsible for Hooker's misinterpretation of the species caption for the plate 3729 of *Curtis's Botanical Magazine,* 1840, as being *Schomburgkia marginata.*

There are five pale yellow, raised veins noticeable on the disc where the lip reflexes, and again, there is a raised swelling toward the base of the lip that is appressed to the column. The latter is a feature described for other species in the genus and must play a role in pollination procedures. It must press the insect proboscis against the stigma and column as it probes for nectar and thus effects pollen transfer.

In this species, the midlobe is about one-half the total length of the lip and is separated from the lateral lobes by short indentations. The midlobe is oval or rounded and somewhat longer than broad. The claw (base) of the lip is attached on either side to the wide and thin wings of the column to a point about one-quarter to one-third of the way from the column base where the free portion begins. The pedicel and ovary are greenish, and the column is white with purple flush on top at the base, more yellow toward the tip and red spotted on the lower surface. Dunsterville and Garay note that there are two red-flushed areas on either side of the anther cap, which is otherwise yellow.

In my examples the flowers are of medium size for the genus. The dorsal sepal is 22 mm long and 8 mm wide at the base, increasing to 12 mm near the tip, which is rounded. The petals are 21 mm long and 6 mm wide at their base and 10 mm at their apex, and again we find the elongated, spatulate shape. The lateral sepals are 8 × 45 mm with less pronounced widening toward their tips. All these segments are either obtuse or rounded at the tips. The lip measures 16 × 20 mm, and pedicel and ovary are about 70–75 mm. The floral bracts are 5–6 cm long, chaffy, and soon quill and droop along the flower stalk.

Nothing seems to have been published on the history of this species. Mostly it has passed its time in confusion, usually labeled either *Schomburgkia crispa* or *S. lueddemanniana* (correctly, *S. lueddemanii*). The next time your plant flowers, look it over carefully. Dunsterville and Garay's examples came from the Perija foothills in the state of Zulia of Venezuela. Consult the drawings in Figure 1-1 for details and proportions of lip structures. Reichenbach wrote that the plant first flowered in the garden of Mr. Booth, cultivated by Mr. Boysen, and that is the total story of the introduction of this species to cultivation.

Galeott's Schomburgkia

Guatemala, México

(PLATE 12)

Schomburgkia galeottiana Richard. 1845. *Ann. Sci. Nat.* s. 3, 3:23.
Subgenus *Chaunoschomburgkia*

Synonyms

Schomburgkia chionodora Reichenbach f. 1886. *Gard. Chron.* 1:73.
Schomburgkia chionodora var. *kimballiana* Reichenbach f. 1888. *Gard. Chron.* 3:136.
Schomburgkia rhionodora var. *kimballiana* Reichenbach f. 1888. *Gard. Chron.*, 3rd s., p. 136. (Misprint for *chionodora*).
Bletia lepidissima Reichenbach f. 1889. *Gard. Chron.* 4:72.
Schomburgkia lepidissima Reichenbach f. 1889. *Gard. Chron.* 3rd s., 1:72.
Schomburgkia sanderiana Rolfe. 1891. *Gard. Chron.* 1:614.
Myrmecophila chionodora (Reichenbach f.) Rolfe. 1917. *Orchid Rev.* 25:51.
**Myrmecophila galeottiana* (Richard) Rolfe. 1917. *Orchid Rev.* 25:51.
Myrmecophila lepidissima (Reichenbach f.) Rolfe. 1917. *Orchid Rev.* 25:51.
Myrmecophila sanderiana (Rolfe) Rolfe. 1917. *Orchid Rev.* 25:51.
Schomburgkia galeottiana var. *kimballiana* (Reichenbach f.) Jones. 1963. *Am. Orch. Soc. Bull.* 32(1):9.
Schomburgkia galeottiana var. *chionodora* (Reichenbach f). Jones. 1968a. *Ceiba* 14:27–37.
Myrmecophila chionodora var. *kimballiana* (Reichenbach f). Kennedy *in litt. Nomen nudum.* 1979. *Orchid Digest* 43(6): 204–211.

This species has been a completely confused concept with a variety of names that may be reduced to synonymy, thus helping clear the schomburgkia air. In contrast to *Schomburgkia exaltata, S. galeottiana* is found on the Pacific side of southern México and adjacent Guatemala.

The flowers are less undulate or crisped on the sepals and petals than other schomburgkias, and the petals are definitely wider than the sepals and stand out nearly at right angles to the dorsal sepal. The sepals show more rippling than do the spatulate petals. Petals measure 11 × 37 mm, the sepals measure 9 × 35 mm. The lip has prominent lateral lobes that turn up at their tips, and the midlobe is expanded widely into a semicircular, fan-shaped, wavy surface with an indented apex. The three, or sometimes five, keeled or raised veins extend from the throat onto the base of the disc. They are neither toothed nor serrate. The sepals and petals are a more or less uniform lavender or purple, somewhat darker on the edges of the petals. The throat of the lip shows white with yellow veining. The tips of the lateral lobes and the midlobe either match the rest of the flower or are a darker magenta-purple. The column is slender for a *Schomburgkia,* and the anther cap shows divergent lobes or, as they are sometimes called, horns.

The good color photos in Wiard's *An Introduction to the Orchids of México* (1987) show the flowers and the plant habit. The pseudobulbs are somewhat more elongate and less conical than those of other species in this subgenus and clearly show the demarcation of five or six nodes. They produce two to four leaves. The elongate flower stalks produce 10–15 flowers in succession. Wiard states "flowering in autumn" in the natural locations in Guerrero, Jalisco, and Michoacan. Hamilton (1986) finds November flowering predominates in cultivation.

The original description of *Schomburgkia chionodora* from dried (faded?) flowers indicates a pale or white-colored form with a purple spot in the center of the lip. Variety *kimballiana* had light purple flowers, and *S. sanderiana,* introduced by Fred Sander and Company, had branching panicles of rosy carmine flowers.

Brieger, Maatsch, and Senghas in *Schlechter III* (1976–1981) believe that this species is the same as *Schomburgkia humboldtii* from Venezuela. They look upon it as an interesting example of disjunct distribution between South and Central America. The Richard name, *S. humboldtii,* would have priority if they are ever considered a single species. Most orchidologists today, however, prefer to consider them as separate entities.

Humboldt's Schomburgkia

Venezuela

(PLATES 13, 14)

Schomburgkia humboldtii (Reichenbach f.) Reichenbach f. 1856. *Xenia Orch.* 1:240; *Bonplandia* 4:328.
Subgenus *Chaunoschomburgkia*

Synonyms

Epidendrum humboldtii Reichenbach f. 1849. *Linnaea, Lit. Bericht* 22:386.
Bletia humboldtii (Reichenbach f.) Reichenbach f. 1861. *Walp. Ann. Bot.* 6:430.
**Myrmecophila humboldtii* (Reichenbach f.) Rolfe. 1917. *Orchid Rev.* 25:51.
Laelia humboldtii (Reichenbach f.) Williams. 1941. *Darwiniana* 5:77.
Schomburgkia humboldtii var. *breviscapa* Jones. 1972a. *Orquideológia* 7(1):7–12.

This species with large flowers was discovered by Humboldt and Bonpland near Puerto Cabello, Venezuela, and was subsequently gathered by Wagener and other collectors in the same locality. In typical schomburgkia fashion, the plants had been in cultivation for years before any flowered. In 1887, flowers were sent from the Backhouse collection in York to Veitch. The plants have never been common in collections, so the species is not well known.

Reichenbach compared the form of the flowers to *Laelia pumila* and the color of the flowers to *L. anceps*. At least one white clone has existed, displayed at a Puerto Rican orchid show I attended, but I find no judging records on it. Jones states (1963) that the albas come from Colombia, though the statement is not confirmed by the herbarium record. *Schomburgkia humboldtii* is the only *Chaunoschomburgkia* to be found in South America, all the others being in México, Central America, or the Caribbean, a distinct disjunction in distribution for the genus. Differences in the lip structure (Fig. 1-1) separate *S. humboldtii* from *S. galeottiana,* although Brieger and his associates consider them as synonymous species.

The pear-shaped, squatty, conical pseudobulbs have two to three leaves and the erect flower stalks may be 1–1.5 m long with up to 20 flowers. Flowers may be as much as 7 cm in natural spread and have purple flushed pedicels. Dunsterville and Garay note that only two or three flowers are open at a time, as the flowers pollinate readily and open successively but stay in prime condition for only a short period. The lilac sepals and petals are undulate with the petals darker colored toward their tips. The lip is a darker purple or amethyst over a white background with the apex a bright purple. The five to seven raised veins are dark purple toward the base of the lip and yellow where they extend onto the disc. The anther cap is white and, like some of the other species, is deeply grooved so that two horns or lobes result.

Hamilton's data indicate March flowering. A natural hybrid, *Schomburgkia parkinsoniana,* based on a single flower (Humboldt #1190) in the Reichenbach Herbarium in Vienna, is presumably between this species and *S. thomsoniana* (var. *minor*) from Cayman Brac Island. One presumes the pollination was carried out by a long-distance bee! The plants of this species, coming from coastal regions, require "much direct sunlight to mature their stems and to insure their flowering regularly," to quote Veitch.

Lueddeman's Schomburgkia

Colombia, Costa Rica, Panama, Venezuela?

(PLATE 15)

Schomburgkia lueddemanii Prilleux. 1862. *Jour. Soc. Imp. Hort..* 8:123.
Subgenus *Schomburgkia*

Synonyms

Bletia undulata var. *costaricana* Reichenbach f. 1865. *Xenia Orch.* 2:49
Laelia lueddemaniana (Prilleux) Williams. 1940. *Ann. Mo. Bot. Gard.* 27:282.
Schomburgkia undulata var. *costaricana* (Reichenbach f.) Jones. 1963. *Orchid Rev.* 71:51.
Schomburgkia undulata var. *lueddemaniana* (Prilleux) Jones. 1963. *Orchid Rev.* 71:51.
Schomburgkia lueddemaniana var. *costaricana* (Reichenbach f.) Jones. 1967a. *Fedde Repert. Spec. Nov.* 76(3):195.

This species is not mentioned in Dunsterville and Garay's *Venezuelan Orchids,* the major authority for that area, nor is it mentioned in Foldats's *Orchidaceae* in *Flora de Venezuela*. *Schomburgkia lueddemanii* was, however, made a synonymn of *S. undulata* when Schlechter published his account of the genus in 1913, though later he restored it to separate species status. The *Index Kewensis* currently lists it as a synonym under *S. undulata.*

Is it a bona fide species or not? Why was it omitted from these modern works when it was listed in 1913 and had been known since 1862? There are plants shown at present under this name in Florida. Are they perhaps misidentified? Do the plants really occur in Venezuela but not in Colombia, or vice versa? What is the actual distribution of *Schomburgkia undulata*? *Schomburgkia lueddemanii* may well be only Central American in distribution, as I don't find any herbarium records for it being in Colombia at all. What is its relationship to *S. wallisii?*

Today the name is properly spelled *Schomburgkia lueddemanii* despite all the variations that have been published. The species was named after a Mr. Lueddeman in Paris, and the type plant was described without its native country or history of collection being listed.

Jones (1970a) says the species is closely allied to *Schomburgkia undulata,* repeating Prilleux's comment of 1862; and in 1973 he further informs us that research shows no type specimen for this species was ever prepared. In discussing the *S. undulata* complex, Jones (1970a) distinguishes carefully among *S. undulata, S. lueddemanii,* and his *S. schlechterana,* identifying them as three separate species. One of the best distinctions for this species is the lip detail. It is ovate to orbicular, apiculate, has a white ground, is shaded with yellow on the disc, is edged with brown, and has three keels. These particular qualities were in Prilleux's original description. Prilleux also said the color of the flowers was an orangey-brown, the sepals were about 5 cm long, the petals were slightly shorter, and all were undulated.

Other authors (Williams, 1946, in particular) have reported the species from Panama, and there are various herbarium specimens attesting this presence. Jones has added another concept to the "luedde" puzzle by including Reichenbach's *Bletia undulata* var. *costaricana* as *Schomburgkia lueddemanii* var. *costaricana* comb. nov. The Costa Rican forms presumably had smaller flowers and were less richly colored than the original species. Costa Rican herbarium material I examined did not bear out these differences, but I have not personally had an opportunity to examine the variation in live plants for comparison. At this point I should say that I do believe that *S. lueddemanii* is a valid concept, though I will reserve opinion on the varietal epithet.

The possible range from Venezuela to Costa Rica, in any case, produces a strange distribution for the species compared to what is known about other orchids of the general area. One may ask why the plants have never been collected in Colombia. There are too many confusions here. A lectotype or neotype specimen should be designated for the species and possibly its variety, and the identifications of all the disputed examples should be rechecked. For instance, the specimen illustrated and described in the 1912 *Curtis's Botanical Magazine,* t. 8427, is not *Schomburgkia lueddemanii* but *S. undulata.*

I do not hesitate to assign to this species the plants labeled *Schomburgkia lueddemanii* that I have seen alive, and that presumably came originally from Venezuela according to the owner (Plate 15). They have been grown and displayed for many years in Florida. The dorsal sepal measures 1.1×5.7 cm, the lateral sepals are 0.9×5.4 cm, the petals are 1.0×4.5 cm, and the flattened lip is 1.4×2 cm. I would describe the flowers as orange-brown in color with blunt tips on the sepals and petals. These parts are not twisted and so are flat in the centers with only the edges rippled in undulations. This is in contrast to the sepals and petals of *S. undulata,* where the whole segment as well as the edges can be twisted around. The lip has a pale ground color in the throat with a central yellow flush and purple edges on the lateral lobes. The midlobe shades to more of a brown on the rim and tip. There is distinct yellow on the disc where the three-keeled veins are located, but they are scarcely raised compared to other species. The lateral veins are more noticeable toward the base, as in the lip drawing (Fig. 1-1), and there is a definite dark spot at the very base of the lip. The pedicel and ovary are brown-green, about 8–9 cm long, and the papery bract is 4.5 cm long. There is a pink flush on the top of the column and a yellow anther cap. The yellow keels and anther cap are noted in the original description.

There is a fine colored illustration of this species in Sheehan and Sheehan's (1968) article under the name of *Schomburgkia undulata*. Note that the short floral bracts are about one-half or two-thirds the length of the pedicellate ovaries and that the bracts and ovaries are white rather than pink or rose-colored like those of *S. undulata*. Brieger and his coauthors in *Schlechter III* consider *S. schultzei* to be a synonym of this species, an idea we will just ignore.

One final complication involves whatever was called *Schomburgkia wallisii*. The records that go with that species name are so scanty and questionable that I have listed it under the heading of "Doubtful Species" at the end of this chapter. Again, there is no type specimen to show us what was actually intended. Rolfe (1913) thought that *S. wallisii* was likely another synonym for *S. lueddemanii*, but until new specimens of genuine Colombian and Venezuelan origins can be compared with those from Central America, I would not like to venture a final opinion.

Lyon's Schomburgkia

Cuba, Jamaica

(PLATE 16)

Schomburgkia lyonsii Lindley. 1853. *Gard. Chron.* 1:615.
Subgenus *Schomburgkia*

Synonyms

Bletia lyonsii (Lindley) Reichenbach f. 1861. *Walp. Ann. Bot.* 6:418.
Schomburgkia carinata Griesbach. 1864. *Flor. B. W. I.*, p. 621.
Laelia lyonsii (Lindley) Williams. 1941. *Darwiniana* 5:76.

Schomburgkia lyonsii is readily distinguished by its white flowers with their rich red-purple spottings and markings; the non-resupinate position of the flowers; the fact that some flowers are incomplete at the top of the inflorescence, lacking petals or having them fused with the dorsal sepal; and that these upper flowers (or sometimes the whole head) are cleistogamous. No other species in the genus resembles this one, and that is a pleasant change from the turmoil that surrounds the identities of some of the other specimens.

The plants were originally found in Jamaica by William Purdie in 1848 and first flowered in cultivation in the collection of J. C. Lyons of Ladiston, England, in 1853. Later they were discovered in Cuba by Wright, an early botanist in the Caribbean. They were found in habitats from sea level to an altitude of about 800 m. Fowlie (1963) found them mostly in forests of drier upland valleys with southern exposures, growing upon rocks in exposed locations. The plants flowered in early fall "in these situations, after maturation of the most recent growth made up from spring rains beginning in mid-May." The Hamilton data show a predominance of July–August flowering in cultivation.

The lip is colored the same as the sepals and petals, also a distinction in this genus, and there are yellow-orange markings on the margins—an interesting contrast. The lateral lobes of the lip are poorly delineated, so that the lip has an elongated triangular shape overall, another unique quality, and there are five raised, central veins. The column is short and strongly incurved from a distinctively thickened base, and the apex displays three teeth from the edge of the clinandrium. There are also the two central linear appendages projecting from the top of the anther cap. These appendages are unique in the genus and should not be confused with the distinct lobing of the anther cap into two sections, sometimes referred to as a horned anther cap.

Measurements of flower parts show the dorsal sepal is 12 × 28 mm; lateral sepals are 13 × 21 mm; petals are 11 × 28 mm; the flattened lip is 12 mm across the base and 15 mm long; the stalked ovary is 50 mm; and scarious, green/pink flushed bracts are 45 mm long. The flower stalk approximates 50 cm. Ordinarily the plants have two leaves about 20–30 cm long on pseudobulbs up to 20 cm high. Flowers number 10–20.

The slightly falcate lateral sepals are comparatively wider than usual, and this can best be explained as an adaptation correlating with these right-side up (non-resupinate) flowers. The lateral sepals are uppermost and appear as petals would in an ordinary species. No pollination studies of this species have been made, nor has its pollinator been noted, but such studies would certainly be welcomed.

A white or alba form of the species has been found in one valley at Trout Hall, according to Fowlie (1963), and Jones (1972b) formally named the white forms as variety *immaculata*. Another alba form was illustrated on the May cover of the *American Orchid Society Bulletin* in 1957. The cover story explains that the plant was collected in 1937 in the hills of the Parish of Clarendon and that they also occurred in two other locations at about 800 m. G. C. Gunter of Half Way Tree, the correspondent sending in the cover illustration, said the racemes of this species were sometimes 1.5 m long with 10–25 flowers measuring 3.8–5 cm inches across and that the large clumps of plants growing in the tops of tall trees, in full exposure to the sun, were difficult to collect.

Yellow-margined Schomburgkia

Brazil, Guyana, Suriname, Venezuela

(PLATES 17, 18)

Schomburgkia marginata Lindley. 1838. *Sert. Orch.*, plate 13.
Subgenus *Schomburgkia*

Synonyms

Cattleya marginata (Lindley) Beer. 1854. *Prakt. Stud. Fam. Orch.*, p. 212.
Bletia marginata (Lindley) Reichenbach f. 1861. *Walp. Ann. Bot.* 6:419.
Laelia marginata (Lindley) Williams. 1941. *Darwiniana* 5:76.

Schomburgkia marginata, the second species of *Schomburgkia* to be named, was found by Schomburgk in Suriname, but Lindley's description of the species was made from plants collected by John Henry Lance during his residence in that country. Lance was an Englishman who served as a judge at the Mixed Court of Justice from 1822 to 1833. For his botanical studies in Suriname he was awarded the Silver Medal of Honour of the Royal Horticultural Society of London in 1834.

Plants of this species, as well as other orchids, were sent to England in 1834 and flowered first in the collection of Thomas Brocklehurst of the Fence, near Macclesfield. In nature, the plants grew commonly near the capital of Paramaribo, always in the first or second main fork of *Erythrina* trees, locally called the coffee-mamma. The orchids flower there from January to March.

The species was named for the narrow yellow edge around the brick-colored sepals and petals. The strongly reflexed lip is white with a pink flush toward the edges and apex and has five raised veins. The ovaries with their stalks are greenish brown, and the columns are white with a flush of pink. Floral bracts are scarious or chaffy and whitish in color. The plants produce two or three leaves and otherwise resemble typical schomburgkias. There is an excellent photo made in Suriname in *Orchids of Suriname* by Marga Werkhoven (1986), but unfortunately it is labeled *Schomburgkia crispa*. Indeed, some orchidists today consider them to be synonymous. In the text she describes the plant as coming from Paramaribo, the type location for the discovery of this species in the early 1830s.

The pseudobulbs are about 10 cm tall, the leaves up to 25 × 6 cm in dimension, and the wand of flowers can be a meter or more in length. Sepals are 2 × 0.9 cm, and petals are 2 × 0.8 cm. The lip, which is strongly recurved, is pale pink or white, sometimes with darker edges, with five light yellow, waved keels. It tapers toward the base and is distinctly three-lobed; the midlobe is pointed. The column is usually flushed with purple and the anther cap is yellow. Flowers appeared in nature during October–December.

This species, as mentioned above, has often been confused with *Schomburgkia crispa,* but the latter lacks the yellow margins on the sepals and petals though it may otherwise have similar configuration. *Schomburgkia marginata* tends toward red-brown color in the sepals and petals, while *S. crispa* has more yellow-brown tones. The color and lip shape ("more acute folds") distinguish it from *S. crispa*. The two species must, however, be closely related, and in turn, *S. marginata* is also confused with *S. fimbriata (gloriosa)*. Flowers of the latter species are also of a red-brown color but have no yellow margins on the sepals or petals. One other species, *S. weberbaueriana,* is likely the same or a congener species with *S. marginata* and could probably be placed in synonymy with it. It is kept separately here because of its separate (disjunct?) distribution in Peru. A painting by Margaret Mee in her excellent book of observations, *In Search of Flowers in the Amazon Jungle* (p. 254, 1988), illustrates *S. marginata* well and vouches for its distribution from Suriname into northern Brazil along the Rio Negro.

Moyobamba Schomburgkia

Bolivia, Peru

(PLATE 19)

Schomburgkia moyobambae Schlechter. 1921. *Fedde Repert. Spec. Nov.* 9:97.
Subgenus *Schomburgkia*

Synonyms

Laelia moyobambae (Schlechter) Schweinfurth. 1944. *Bot. Mus. Leaf., Harvard Univ.* 11:180.

Schomburgkia undulata var. *moyobambae* (Schlechter) Jones. 1963. *Orchid Rev.* 71:51.

Schomburgkia marginata var. *moyobambae* (Schlechter) Jones. 1968. *Lloydia* 31:269.

The plant habit of *Schomburgkia moyobambae* is the same as for other fusiform pseudobulbed species, and the flower stalk reaches 30 cm or more with a dense cluster of 20–25 brownish flowers that can have yellowish margins on the sepals. The rose-colored bracts, about two-thirds as long as the pedicellate ovaries, produce an interesting contrast with the yellow-brown of the flowers. The sepals, about 23 mm long, are blunt, almost rounded, at the tips; the petals, about 20 mm long, come to a point. The segments have the usual crisped or undulate margins. The column and lip are rose-colored, matching the bracts, and the lip is strongly recurved along its length, with the edges raised. It is 15 mm long and 9 mm wide. There are five keeled veins over the disc alternating with unkeeled veins. The lip is three-lobed; the lateral lobes are short and broad with blunt triangular apices; the midlobe is also short, broadly oval and blunt or rounded at the tip. The column is 13 mm long and dilated at the stigma. The pedicellate ovary is 55 mm long.

Little seems to have been written about *Schomburgkia moyobambae* though it appears in an unusual location, sharing the honor with closely related *S. weberbaueriana*. They are the two southernmost species in the whole genus. *S. moyobambae* was first found in Loreto, near Moyobamba, Peru, by Dr. Serafin Filomeno. Because of the yellowish margin on the sepals of the red-brown flowers, it has sometimes been confused with *S. marginata,* though this marginal pattern is not too pronounced in some clones. The length and color of the bracts can immediately distinguish this species from *S. marginata* even if the country of origin is not known at the time. The rose-colored lip of *S. moyobambae,* contrasting with the white or pale pink lip of the *S. marginata,* provides another difference; and it has smaller flowers and a shorter and smaller lip with smaller projecting side lobes than *S. weberbaueriana*.

I have not been able to find out much more about this species. More information would be helpful. See the discussion under *Schomburgkia marginata*.

Rosy Schomburgkia

Colombia, Venezuela

(FIGURE 1-2; PLATE 20)

Schomburgkia rosea Linden *ex* Lindley. 1845. *Bot. Reg.* 31, sub t. 53.
Subgenus *Schomburgkia*

Synonyms

Bletia rosea (Linden *ex* Lindley) Reichenbach f. 1861. *Walp. Ann. Bot.* 6:420.

Laelia rosea (Linden *ex* Lindley) Schweinfurth. 1967. *Mem. N.Y. Bot. Gard.* 14(3):155.

The plants of *Schomburgkia rosea* have typical *Schomburgkia* qualities, but the flowers are among the smallest in the genus, sharing that distinction with *S. wendlandii*. The flowers are also distinctive in their dark rose-red or maroon color with the floral bracts a paler rose-red. The ovaries are a pale pink, and the flower stalk, where it is not covered with bracts, is a dark brown-purple. An examination of the light pink lip shows that the side lobes are each nearly as large as the midlobe. The column is rose-purple with fine spottings of darker purple. The colors are retained on drying and do not turn black as so many herbarium specimens of this genus are prone to do. It does not seem ever to have been a common plant in cultivation.

The dorsal sepal measures 6 × 25 mm, and the petals are 5 × 23 mm. The lip flattened is 13 mm wide and 19 mm long in my flower specimen from a plant observed at a show in Caracas many years ago. It can be nearly circular in outline with the midlobe smaller than the lateral lobes. Some purple spotting can occur along veins toward the base of the lip, but the forward portions of the keels are white. The three central and two lateral keels are most prominent, and two adjacent to the center keel are smaller, making seven in all. There can be 10–28 flowers, all in a compact, crisped, and undulating group. The pseudobulbs reach 20 cm, and the stiff leaves may measure 5 × 27 cm and are noticeably folded along the midrib. The flower stalk may reach 35 cm.

Little is known of the habitat of *Schomburgkia rosea* other than that it is from low mountains and is, therefore, somewhat cooler growing than *S. crispa*. Dunsterville and Garay cite it as coming from near Cajmera, Perija, in Zulia State, Venezuela, and Schweinfurth's (1967) account of the orchids of the Guayana Highland document its presence there. The type specimen was collected in Colombia, and Reichenbach tells us it was found growing on rocks on the northern slope of the Sierra Nevada de Santa Marta, where it was "5000 ft above sea level, between the villages of San Antonio and San Miguel in the territory of the Auruhuacos Indians." Hamilton's data indicate a March flowering in cultivation, and that matches my observations on this species in Venezuela. Brieger and coauthors (1976–1981) make *S. splendens* (do they mean *splendida*?) a synonym of this species, but it has a different callus and much larger flowers that are easily distinguished from those of *S. rosea*.

There is a natural hybrid of *Schomburgkia rosea* × *S. undulata* that was named *S. heidii* as a species before its hybrid nature was understood. It was discovered by Dr. Germán Carnevali, a professor of botany at the Universidade Central de Venezuela in Maracay. He is working now at the Missouri Botanical Garden. The name honors Heidi Pivat, who provided the illustrations for the publication of the type description (Carnevali, 1982). Two plants were found growing on slabs of granite rock and flowering in February in the Territorio Federál Amazonas. The region borders the Orinoco River at a point north of Puerto Ayacucho.

The flowers of *Schomburgkia heidii* are most similar to the *S. rosea* parent in the deep purple-brown color of the sepals and petals, and the rose-purple ovaries and lips. The bracts are violet. It shows evidence of the *S. undulata* parentage by having a modified callus and more elongated sepals, petals, and lip than the typical *S. rosea* flower.

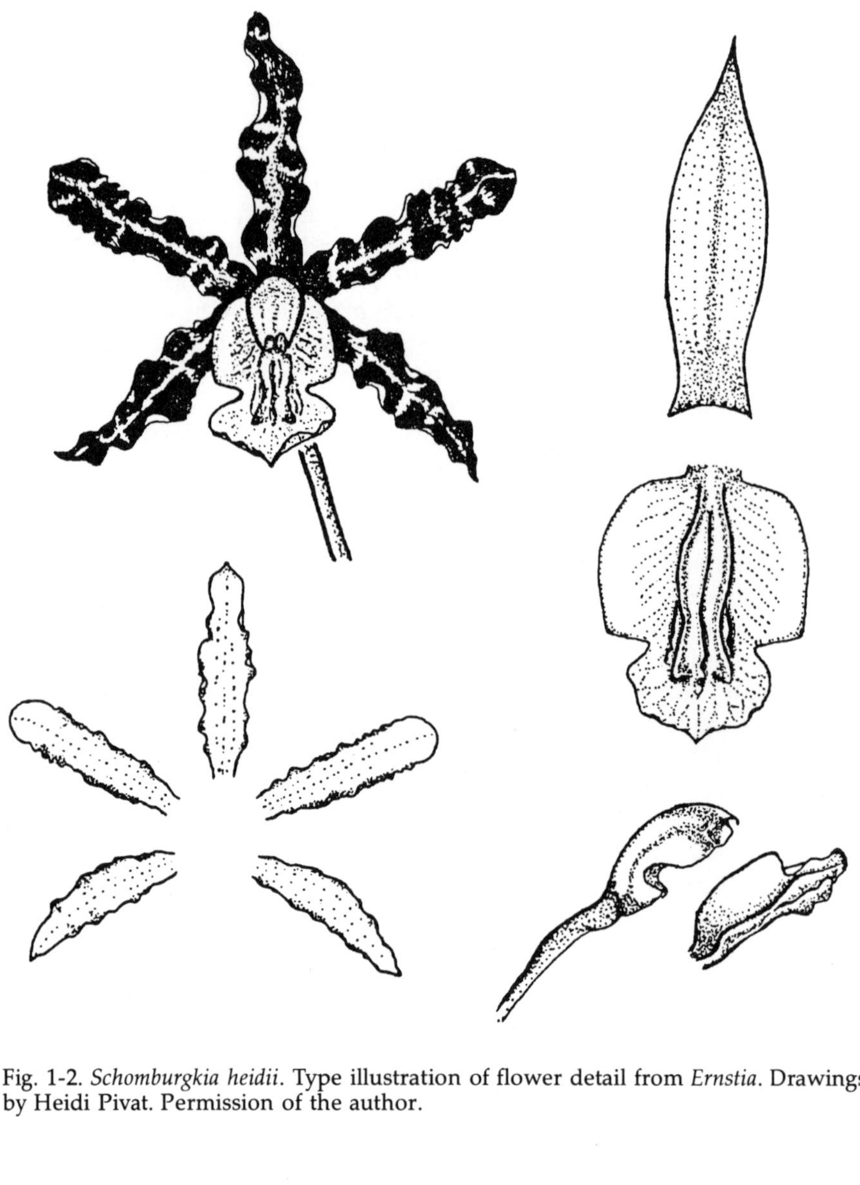

Fig. 1-2. *Schomburgkia heidii*. Type illustration of flower detail from *Ernstia*. Drawings by Heidi Pivat. Permission of the author.

Sawyer's Schomburgkia

México

(PLATE 21)

Schomburgkia sawyeri (Williams) Withner. 1993. *Comb. nov.*
Basionym: *Laelia sawyeri* Williams. 1943. *Am. Orch. Soc. Bull.* 11(9):329.
Subgenus *Chaunoschomburgkia*

So few *Schomburgkia sawyeri* plants have apparently been brought into cultivation or studied in the field that a debate exists over whether they represent a valid species or not. The two live plants I have studied—one at Santa Barbara Orchid Estates, the other in my own collection—certainly have different flowers than those of *S. galeottiana* with which it has usually been confused or made synonymous (see Brieger, Maatsch, and Senghas, 1978). In 1942 Sawyer collected the original plant at about 1500 m near Acuitlapan, Guerrero, a mountainous area on the Pacific slope of México. The limited distribution of this species extends across the state border into an area around Alpuyica, Morelos. The plants are epiphytic on acacia trees.

The flowers are drooping at the end of a nodding stalk and 3–5 to 10–20 may be produced. The raceme may be 1.5 m long. The pseudobulbs are two leaved or occasionally have one leaf. The flowers are magenta-purple overall; the throat of the lip is white with lilac veining and a faint yellow disc. The column is dark violet. The sepals and petals are undulate but not curly edged, and the obovate-oblanceolate petals are wider than the dorsal sepal. The midlobe has serrated and crisped edges that are indented, almost lobed, and is more or less square in outline. The lateral lobes take up half or more of the lip length and have broadly rounded tips. There are 3–5 raised veins or ridges extending from the base of the lip to the disc of the midlobe, the central vein being longer than those on either side.

My plant flowers during July and August. The sepal and petal shape and the midlobe configuration indicate, in my opinion, an affinity with *Schomburgkia superbiens*. This species would appear to lie somewhere between that species and *S. galeottiana* in its characteristics. More information about it is needed.

Schultze's Schomburgkia

Colombia

(PLATES 22, 23)

Schomburgkia schultzei Schlechter. 1924. *Fedde Repert. Spec. Nov.* 27:139.
Subgenus *Schomburgkia*

Synonym

Schomburgkia rosea var. *schultzii* (Schlechter) Jones. 1970b. *Caldasia* 10(50):493.

Schomburgkia schultzei is endemic to Colombia. It is well illustrated in Kennedy's paper (1979). He collected specimens on bare rocks near Pasto in the mountains of southwestern Colombia near the Ecuadorian border. The flowering type specimen was collected by Schultze in March, 1922, from rocks in the Rio Cabrera valley near Cundinamarca, Colombia, at 1600 m. Schlechter wrote that the species is distinguished by short bifoliate pseudobulbs only 15–17 cm high, by the proportionately short lip, and by the porcelain-like shine.

As an examination of the illustration will show, the large, bold flowers are definitely bow-legged compared to other schomburgkia flowers, and they do have wide segments with a high shine. The lateral lobes of the lip do not close over the column as they do in *Schomburgkia splendida,* nor is there a prominent yellow spot on the lip. There are three or more main veins on the disc, but they are not keeled and are barely raised over the lip surface, forming a kind of noticeable prominence that extends to a point at the apex of the lip.

The sepals and petals are a rich reddish brown that contrasts nicely with the magenta lip. The dorsal sepal measures to 35–55 mm long and lateral sepals are 32–50 mm. The lip is adnate to the column for about 2.5 mm. The edges of the lateral lobes and the apex of the midlobe are a darker color. There is lighter color, even white, in the throat and there may be some pale yellow flushes on the disc as the dark veining extends from the throat out onto the midlobe. The lip is 16–43 mm in length. The column is a purple-magenta color and is 10–34 mm long with a white or creamy anther cap. The pedicellate ovary of 45–85 mm is matched in length by the broad, rosy colored bracts that are clustered tightly at the base of the flowers. The measurements show that the flowers vary considerably in size, and this in turn reflects the size and condition of the plant and the number of flowers being produced. Also, some clones may be larger flowered than others.

Although some of these plants are grown in California—probably as a result of the specimens brought back by Kennedy, Fowlie, and Spencer—this species is poorly known in cultivation and is worthy of greater attention for its bold flowers. It has been confused with *Schomburgkia splendida* and also *S. elata,* so that we must be on the lookout for properly labeled plants in collections and shows. Jones reduces this species along with *S. rosea* to synonymy under *S. undulata,* both obviously misunderstandings of its rightful status.

PLATE 1. *Schomburgkia albopurpurea*. Flowering habit. Plant from Cayman Island. Photo by George Kennedy; courtesy of Ruth Kennedy.

PLATE 2. *Schomburgkia albopurpurea*. Flower detail. Photo by George Kennedy; courtesy of Ruth Kennedy.

PLATE 3. *Schomburgkia brysiana* in Belize. Photo by Chris Turnbull.

PLATE 4. *Schomburgkia brysiana*. Flower detail.

PLATE 5. *Schomburgkia crispa*. Type illustration from *Sertum Orchidaceum*, plate X (Lindley, 1838).

PLATE 6. *Schomburgkia crispa*. Flower cluster.

PLATE 7. *Schomburgkia elata*. Flower cluster. Photo by Trudi Marsh.

PLATE 8. *Schomburgkia elata*. Lip detail. Photo by George Kennedy; courtesy of Ruth Kennedy.

PLATE 9. *Schomburgkia exaltata*. Photo by Otto Tinschert.

PLATE 10. *Schomburgkia fimbriata* in Rondônia, Brazil. Flower detail. Photo by Lou C. Menezes.

PLATE 11. *Schomburgkia fimbriata* in *Curtis's Botanical Magazine*, plate 3729 (1840). Published originally as *Schomburgkia marginata*.

PLATE 12. *Schomburgkia galeottiana*. Flower detail.

PLATE 13. *Schomburgkia humboldtii*. Flower detail.

PLATE 14. *Schomburgkia humboldtii,* alba-form. Flower detail. Photo by Jack A. Fowlie.

PLATE 15. *Schomburgkia lueddemanii*. Flower detail.

PLATE 16. *Schomburgkia lyonsii*. Flower in normal non-resupinate position. Photo by George Kennedy; courtesy of Ruth Kennedy.

PLATE 17. *Schomburgkia marginata*. Type illustration from *Sertum Orchidaceum*, plate XIII (Lindley, 1838).

PLATE 18. *Schomburgkia marginata*. Flower cluster.

PLATE 19. *Schomburgkia moyobambae*. Flower cluster.

PLATE 20. *Schomburgkia rosea*. Flower cluster.

PLATE 21. *Schomburgkia sawyeri*. Flower habit.

PLATE 22. *Schomburgkia schultzei*. Flower cluster. Photo by Robert Hamilton.

PLATE 23. *Schomburgkia schultzei*. Flower detail.

PLATE 24. *Schomburgkia splendida*. Flower detail.

PLATE 25. *Schomburgkia splendida*. Water color of *S. imperialis* prepared by Matilda Smith for *Curtis's Botanical Magazine* but never published. Courtesy of Royal Botanic Gardens, Kew.

PLATE 26. *Schomburgkia superbiens*. Flower cluster.

PLATE 27. *Schomburgkia superbiens*. Lip detail. Photo by George Kennedy; courtesy of Ruth Kennedy.

PLATE 28. *Schomburgkia thomsoniana* from Cayman Brac Island. *Curtis's Botanical Magazine,* t. 7815 (1902).

PLATE 29. *Schomburgkia thomsoniana*. Flower detail. Photo by George Kennedy; courtesy of Ruth Kennedy.

PLATE 30. *Schomburgkia thomsoniana* 'Nono' CHM/AOS.
Unusual xanthotic color form. Plant belongs to Zelo Wride. Courtesy of the American Orchid Society.

PLATE 31. *Schomburgkia tibicinis*. Illustration from *Orchidaceae of México and Guatemala,* t. 30 (Bateman, 1843).

PLATE 32. *Schomburgkia tibicinis*. Flower detail. Courtesy of the American Orchid Society.

PLATE 33. *Schomburgkia undulata*. Flower cluster.

PLATE 34. *Schomburgkia undulata*. Type illustration from *Pescatorea,* 1860. Photo courtesy of Rudolf Jenny.

PLATE 35. *Schomburgkia weberbaueriana* 'Robert' CHM/AOS, CCM/AOS. Specimen plant. Grower Robert Fuchs. Courtesy of the American Orchid Society.

PLATE 36. *Schomburgkia weberbaueriana* 'Robert' CHM/AOS, CCM/AOS. Flower detail. Grower Robert Fuchs. Courtesy of the American Orchid Society.

PLATE 37. *Schomburgkia wendlandii*. Non-resupinate flower habit. Photo by Rebecca Northen.

PLATE 38. *Schomburgkia wendlandii*. Flower detail.

PLATE 39. *Schomburgkia ?schlechterana* 'Verde'. Venezuela.

PLATE 39A. *Schomburgkia ?wallisii.* Flower detail. Photo courtesy Carlos Garcia Esquivel, Colombia.

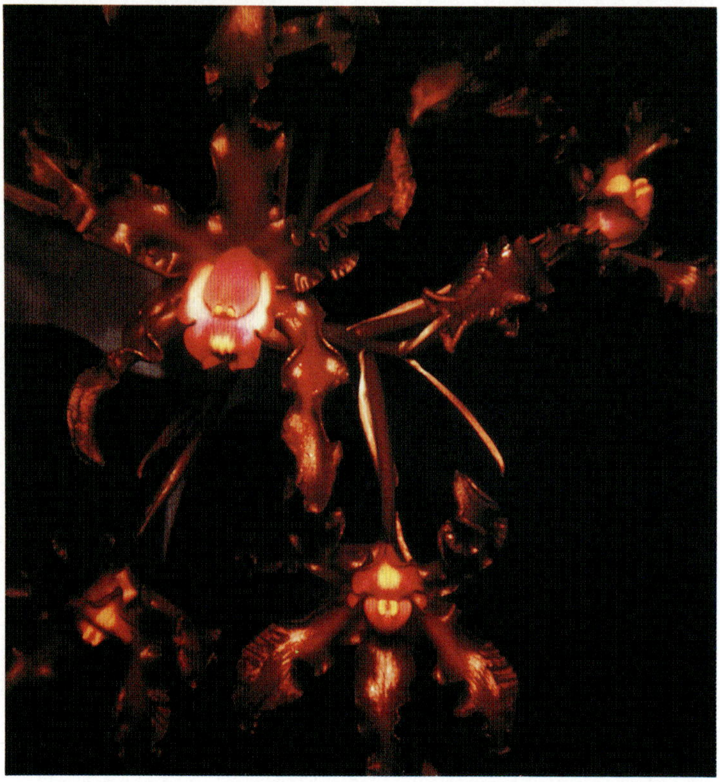

PLATE 40. *Sophronitis acuensis*. Flower and plant detail.

PLATE 41. *Sophronitis bicolor*. Flower and plant detail.

PLATE 42. *Sophronitis brevipedunculata*. Note the habit of this species, the naturally good shape of the flower, and the usual good red coloration. Photo by Jorge Verboonen.

PLATE 43. *Sophronitis brevipedunculata* 'Ericka' nominated for an AM/AOS, Eastern Orchid Congress, 1967, and identified at that time as *S. rossiteriana*. Note the unusual salmon-tinted tones of the flowers.

PLATE 44. *Sophronitis cernua*. Plant habit.

PLATE 45. *Sophronitis cernua*. Flower detail.

PLATE 46. *Sophronitis coccinea*. Flower and plant detail. From *Flore des Serres* (1868). Courtesy of the Royal Botanic Gardens, Kew.

PLATE 47. *Sophronitis coccinea*. Type illustration by Descourtilz along with *Cattleya bicolor* from *Sertum Orchidaceum*, plate V (Lindley, 1838).

PLATE 48. *Sophronitis coccinea*. Plant producing two flowers per growth. Photo by Irene Rohrmoser.

PLATE 49. *Sophronitis coccinea* var. *flava*. Photo by Jorge Verboonen.

PLATE 50. *Sophronitis coccinea* var. *lowii* AM/RHS. Nellie Roberts's painting of the awarded clone, 16 January 1923. Plant of Pantia Ralli. Courtesy of the Royal Horticultural Society, Orchid Committee, London.

PLATE 51. *Sophronitis coccinea* var. *rossiteriana*. Nellie Roberts's painting of the AM awarded clone, 10 January 1899. Plant of Sir Trevor Lawrence. Courtesy of the Royal Horticultural Society, Orchid Committee, London.

PLATE 52. *Sophronitis mantiqueira* by Margaret Mee. Permission by arrangement with Greville Mee.

PLATE 53. *Sophronitis mantiqueirae* plants for sale at Floralia nursery, Petrópolis, Brazil, 1989.

PLATE 54. *Sophronitis mantiqueirae*. Flower and plant variation at Orquidário Binot nursery, Petropolis, Brazil. Note that the lips have little yellow pigmentation.

PLATE 55. *Sophronitis mantiqueirae* var. *xanthocheila* at Floralia nursery, Petrópolis, Brazil. Note the yellow pigmentation of the lips.

PLATE 56. *Sophronitis pterocarpa*. Photo by Denis Duveen. Courtesy of Jorge Verboonen.

PLATE 57. *Sophronitis pygmaea* 'Mignon'. Flower and plant detail. Photo by Roberto Kautsky. Courtesy of Milton Leinig.

PLATE 58. *Sophronitis wittigiana*. Flower detail.

PLATE 59. *Sophronitis wittigiana*. Type illustration for *Sophronitis grandiflora* var. *rosea* (synonym) from *The Garden* (1884).

PLATE 60. *Constantia cipoensis*. Flower and plants. Photo by Trudi Marsh.

PLATE 61. *Constantia cristinae*. Photo by Francisco Miranda.

PLATE 62. *Constantia microscopica*. Photo by Francisco Miranda.

PLATE 63. *Constantia rupestris*. Copy of Rodrigues's original painting by Miss Dyer. Also used as plate 119-II in Martius 3(5). Courtesy of Royal Botanic Gardens, Kew.

PLATE 64. *Isabelia virginalis*. Plant showing "woven" bracts. Photo by Gary Roedell.

PLATE 65. *Isabelia virginalis*. Flower detail. Photo by Rudolf Jenny.

PLATE 66. *Lanium avicula*. Photo by Trudi Marsh.

PLATE 67. *Lanium avicula*. Flower detail with pubescence on stalks and sepals. Photo by Rudolf Jenny.

PLATE 68. *Leptotes bicolor*. Flower detail. Photo by Trudi Marsh.

PLATE 69. *Leptotes bicolor* var. *alba*. Plant and flower detail.

PLATE 70. *Leptotes pauloensis*. Type illustration by Hoehne (1933).

PLATE 71. *Leptotes pauloensis.*
Flower detail. Photo courtesy of Larry Moscowitz.

PLATE 72. *Leptotes serrulata.* Type illustration from *Sertum Orchidaceum,* plate XI (Lindley, 1838).

PLATE 73. *Leptotes tenuis*. Flower and plant detail. Photo by Trudi Marsh.

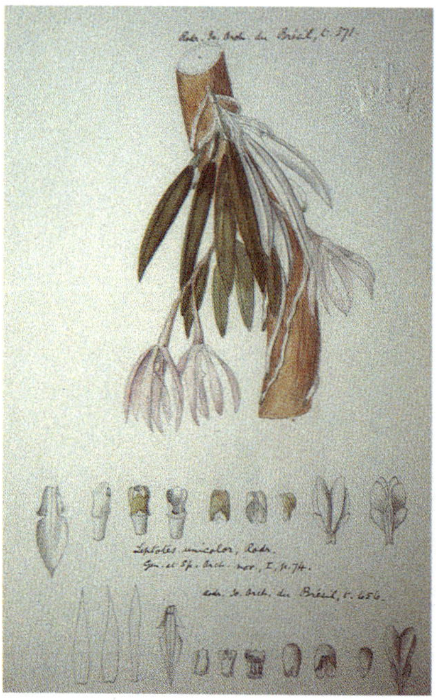

PLATE 74. *Leptotes unicolor* and *L. paranaensis*. Plate 59 from Martius 3(5). Copy of Rodrigues's original painting by Miss Dyer. Courtesy of Royal Botanic Gardens, Kew.

PLATE 75. *Leptotes unicolor*. Flower detail. Photo by Trudi Marsh.

PLATE 76. *Loefgrenianthus blanche-amesiae*. Watercolor by Lofgren from Hoehne's *Iconografia* (1949). Permission of the Instituto de Botanica, São Paulo, Brazil.

PLATE 77. *Neolauchea pulchella*. Flower detail. An unusually dark form. Photo by Rudolf Jenny.

PLATE 78. *Pseudolaelia corcovadensis* 'Karen' CBR/AOS. Courtesy of the American Orchid Society.

PLATE 79. *Pseudolaelia dutrae*. Photo by Euclidio Colnago. Courtesy of Milton Leinig.

PLATE 80. *Pseudolaelia vellozicola*. Photo by Euclidio Colnago. Courtesy of Milton Leinig.

PLATE 81. *Pygmaeorchis seidelii*. Preserved type specimen. Photo by Antonio Toscano.

PLATE 82. *Reichenbachanthus* species. Plant habit. Photo by D. and T. Mulder (1990).

PLATE 83. *Reichenbachanthus* species. Flower detail. Photo by D. and T. Mulder (1990).

PLATE 84. *Renata canaanensis*. Augusto Ruschi examining a *Renata* plant. From Ruschi (1986). Permission of the publisher, Expressão e Cultura, Rio de Janeiro, Brazil.

PLATE 85. *Sophronitella violacea*. Plant and flower detail.

Splendid Schomburgkia

Colombia, Ecuador

(PLATES 24, 25)

Schomburgkia splendida Schlechter. 1913. *Fedde Repert. Spec. Nov.* 12: 212.
Subgenus *Schomburgkia*

Synonym

Laelia splendida (Schlechter) Williams. 1941. *Darwiniana* 5:77.

There is so little printed about *Schomburgkia splendida* that it is difficult to find references to it in the literature, even in Schlechter's own writings on the genus. It is probably the most beautiful species in the genus, with its varnished, deep, almost black, wine-red flowers and large, nearly petal-like pink bracts. The ovary is also rose-purple. The lip is a contrasting darker magenta with a yellow or creamy area on the disc. A good head of these oversize blooms produces an outstanding sight. The plants in nature grow in sunny, exposed locations in low forests along riverways and may be either on the trees or rocks. Herbarium material shows it to occur in southwestern Colombia or in northern Ecuador, and the type was collected at Boqueron del Dagua, Colombia, in the Western Cordillera at 1000 m. Some authors have misspelled the epithet *S. splendens.*

The growth habit is typical, having stout fusiform pseudobulbs covered with tight bracts and two or three long, rather narrow leaves. The flower stalk may reach a meter or more and produces 10–15 flowers, second in size only to *Schomburgkia superbiens*. The dorsal sepal measures from 11 × 50 mm up to 17 × 54 mm; the pedecillate ovary measures to 85 mm; and the rosy pink floral bracts reach 95 mm, longer than the pedicellate ovary. The petals are oblanceolate, widest across their tips but shorter than the dorsal sepal. When flattened, the lip measures 28 mm across the lateral lobes and about 30 mm long in my specimen and 25 mm across × 40 mm long in the type. The side lobes are well defined and pronounced, curling up and touching over the column, though not completely covering it. The throat is pale yellow overlaid with magenta, at the lip's point of maximum convexity the disc is yellow. The lip is much arched when seen in profile, and the five white keels are easily seen over the raised convexity of the disc. The column itself is round and fleshy, rose-purple from a greenish base, and the anther cap is yellow with two lobes or protrusions separated by the central tooth of the column. The bracts measure 10 mm wide and up to 80 mm long.

The flowers are nicely illustrated with a line drawing in the Dodsons' *Icones* of Ecuadorian orchids (1980), plate 487, but otherwise seem to have escaped recent illustration in journals except for Kennedy's paper (1979) and the cover of *Orquideologia* (Colombia) for March 1972.

A selected clone should be pollinated or sib-crossed so that the species could be more readily available for its handsome display. No one has really explored the hybrid possibilities of this group nor of this species especially. It should combine well with *Cattleya* or *Laelia* to produce clusters of dark-colored flowers on smaller, tolerant, easier-to-manage plants. One might guess that the color inheritance could follow the same patterns as hybrids with *L. tenebrosa,* which was much used in early *Cattleya* hybridizations to intensify colors and eventually produced near reds. The genes of this species might do even better on the color, and would also not decrease the flower size or plant vigor in the hybrids.

Superb Schomburgkia

Guatemala, Honduras, México

(PLATES 26, 27)

Schomburgkia superbiens (Lindley) Rolfe. 1917. *Orchid Rev*. 25:49.
Subgenus *Schomburgkia*

Synonyms

Laelia superbiens Lindley. 1840. *Bot. Reg*. 26: misc., p. 46.
Cattleya superbiens (Lindley) Beer. 1854. *Prakt. Stud. Fam. Orch*., p. 214.
Bletia superbiens (Lindley) Reichenbach f. 1862. *Walp. Ann. Bot*. 6:418.

This species, known successively as a *Laelia, Cattleya,* and *Bletia,* was finally recognized for what it was and transferred, after 77 years, to *Schomburgkia* in 1917 by Rolfe. He notes in his article that there were two fine specimens in flower at Kew and that

> careful comparison leaves no doubt that the species belongs to *Schomburgkia,* the fusiform, two-leaved pseudobulbs, the long scapes with numerous elongate bracts, and the details of the flowers, with the numerous undulate keels on the disc of the lip, being all in agreement.

As is described in my introductory remarks, Lindley, who founded the genus, made a mistake in not classifying this species as a *Schomburgkia*. He did not realize it was one of his own children. Then Schlechter, writing up the first monograph on the group (1913), also missed it, so it is easy to understand why so many have been confused as to whether to call it a *Laelia* or *Schomburgkia*. There is no question at this point that it has greater affinity with the latter taxon.

The plants were originally discovered by George Ure Skinner in 1839, planted in front of some Indian settlements and in the *barrancos* of Sachmarachon near Comalapa:

> Saw you anything like this! . . . It is decidedly Epidendrea, but I expect it will get a new title in Europe. It flowers in November, and in some instances bears from 18–20 flowers on stems from 9–12 ft. [3–4 m] long!

They were then found in the wild about 60 kilometers north of Guatemala City, near the village of San Martín Jilotepeque. I had the good fortune to visit the forests of that area in 1986, and, as Rolfe says

> the plants exist in enormous quantities, the finest specimens growing out of rocks, where it is sheltered from the north wind, some of the spikes being 12 feet [4 m] long and bearing upwards of twenty light purple flowers. Owing to this character it has been called the Wand of St. Joseph.

In Spanish, the plant is called *la vera del Señor San José*. The specimens are indeed superb and imposing, clumps exceeding a meter in diameter, some forming complete wreaths around the lower forks of trees with individual pseudobulbs to 50 cm high. This plant does not make a good subject for growing under lights in a confined environment!

The pseudobulbs may reach 35 cm height and there are 1–2 very stiff, leathery leaves up to 6.5 × 30 cm. Large papery sheaths cover the pseudobulbs and the flower stalk as well. The dorsal sepal on my specimen measures 1.5 × 8 cm; the petals are 1.7 × 8 cm; and the flattened lip is 3.2 cm wide across the lateral lobes and 4.8 cm from base to apex. The pedicel and ovary are 6 cm long. Note that the petals are a little wider than the sepals in this species and that the amount of crisping and undulation of the sepals and petal margins is minimal for the genus.

The sepals and petals are purple, varying from paler to darker, and the lip has darker purple veining, with the three central, prominently yellow-crested central veins decorating the disc of the midlobe. The crested lateral veins are usually yellow, but can be dark purple. Sometimes the edges of the midlobe isthmus are minimal in width and reflex back to such a degree that only the cluster of keeled veins and their tooth-like projections are apparent between the throat and the apical portion of the midlobe. At other times the isthmus is more noticeable with the yellow teeth and keels not so obviously displayed as a separate portion of the flower. The ovaries are green flushed with purple, and the 12 cm floral bracts are papery and turn brown. The column is flushed with purple, and the anther caps are a contrasting white.

The taxon was introduced to cultivation in 1842 by Hartweg, who sent plants to the Horticultural Society of London. They flowered for the first time in February 1844 in the collection of Mrs. Wray, of Oakfield, near Cheltenham, England. As Rolfe said, "a noble species, but a little too large for many ordinary houses." My plant has flowered more readily than some of the other species in the genus, perhaps because of its forest habitat it does not demand the most intense light that is necessary for the other species to bloom properly.

A rare white form called 'Don Hubert Cross' has been brought into cultivation and is in a few Florida collections. It was granted a CBM/AOS of 82 points in 1970 and is listed as *Laelia superbiens* instead of *Schomburgkia*, which we now know should have been its generic epithet. *Don* is a title, not a given name.

Thomson's Dwarf Schomburgkia

Little Cayman and Cayman Brac Islands, Cuba

(PLATES 28, 29, 30)

Schomburgkia thomsoniana Reichenbach f. 1887. *Gard. Chron.* 3rd s., 2:38, and 1891, *l.c.* 1:615. New lectotype is Fawcett, 1888, Jamaica, at the New York Botanical Garden Herbarium.
Subgenus *Chaunoschomburgkia*

Synonyms

Bletia thomsonia (Reichenbach f.) Reichenbach f. 1887. *Gard. Chron.* 3rd s., 2:38.
Schomburgkia thomsoniana var. *minor* Strachan *ex* Fawcett. 1894. *Bull. Bot. Dept., Jamaica,* n.s., 1:74.
Schomburgkia thomsoniana var. *minor* Hooker f. 1902. *Curtis's Bot. Mag.,* t. 7815.
**Myrmecophila thomsoniana* (Reichenbach f.) Rolfe. 1917. *Orchid Rev.* 25:51.
Laelia thomsoniana (Reichenbach f.) Williams. 1941. *Darwiniana* 5:77.
Schomburgkia brysiana var. *thomsoniana* (Reichenbach f.) Jones. 1963. *Am. Orch. Soc. Bull.* 32(1):5–12.
Schomburgkia brysiana var. *minor* (Hooker f.) Jones. 1966. *Ann. Naturhist. Mus. Wien* 69:57–67.

Schomburgkia thomsoniana has been through the taxonomic mill. It has been transferred to another species and to all likely genera of this subtribe but *Epidendrum.* It has finally settled, we hope, into its proper designation with the publication of this description.

When Reichenbach first introduced and described the plant, it had flowered in the collection of W. J. Thomson (there is no *p* in the name) at St. Helen's, Lancashire. He was the former Conservator of Cinchona Plantations in Jamaica, and first displayed the *Schomburgkia thomsoniana* at a horticultural exposition in Liverpool.

Jones (1976) notes that no original type specimen was prepared at the time, a situation that often prevailed in the early days of orchid growing before the type system became official, with rules and regulations for all botanists and horticulturists. There is a Fawcett specimen, dated 1888, at the New York Botanical Garden (and three duplicate specimens of the same collection are in the Kew Herbarium) that I would consider a lectotype specimen, since the date coincides with Strachan's discovery of the species' habitat (*Bull. Bot. Dept., Jamaica,* 1894) and the sending of live plants to Kew.

The type description was published in 1887 from a plant that was awarded a certificate of merit in London on July 2 to Sir Trevor Lawrence of Burford. The *Curtis's Botanical Magazine* plate was prepared from this Lawrence specimen. In 1894 the species had been divided into two varieties, var. *albopurpurea* and var. *minor.* (See the discussion under *Schomburgkia albopurpurea,* since that variety has been raised to species status in this book.) According to the rules of nomenclature, the variety *minor* retains the original name of the concept when raised to species status. It is now included in the epithet and becomes a nomenclatural synonym under the *S. thomsoniana.* (The reader will find that the descriptions of var. *minor* directly match what was called *S. thomsoniana* in previous accounts.)

The account in Veitch's *Manual* is based on the 1887 description and states that the leaves and pseudobulbs were like those of *Schomburgkia tibicinis,* but smaller. For this genus, in fact, it may be thought of as a dwarf species, and *S. thomsoniana* has been used in several hybrids because of its comparatively small growth habit as well as its colorful flowers.

The sepals and petals measure about 6–7 × 25–30 mm. The petals are a little narrower than the sepals and have greater crisping of their margins. Flowers have a natural spread of 5–6.7 cm and are of a fleshy substance. The sepals and petals are buff to bright yellow, the clarity of the yellow varying from plant to plant, often appearing dull or dirty looking. Naturally, growers and hybridizers look for clones with bright, clear colors. Kennedy (1979) illustrated this species well but without a proper name, noting only that it came from Cayman Brac Island. It also occurs in Cuba.

The lip is white near the base and is marked with central yellow veins and lateral maroon-purple veining. The midlobe is deep maroon-purple, almost a red in some forms, and the lateral lobes are white on their outside surfaces as they curl and wrap over the pale green column with lavender markings. The much-crisped, almost unopened looking midlobe has a tendency to reflex back and under as a result of the dark granulations and verrucosities over the veins at the front edge, but this characteristic is not pronounced in certain clones. The disc is yellow and white with five raised dark yellow veins, the outer and central larger than the two enclosed. These features are strongly passed on to hybrids. The flowers sometimes have sepals and petals that range from white or near white to yellow, but these may be the natural hybrid forms from Grand Cayman Island and not the pure species. As many as 12–20 flowers on a short, sometimes branched stalk are produced in succession.

Plants in cultivation flower in the early summer, with July the peak according to Hamilton's data. They do best in hanging baskets potted in osmunda fiber placed close to the glass with a maximum of warmth and light. They grow on scrub exposed to full sun and salt spray in their native environment.

The discussion under *Schomburgkia albopurpurea* details the natural hybrids between *S. albopurpurea* and *S. thomsoniana,* called *S.* ×*intermedia,* which occur on Grand Caymen as mentioned above. The species called *S. campecheana* by Kraenzlin in 1903 was thought to be a natural hybrid of *S. thomsoniana* × *S. undulata,* but this is unlikely in terms of our present knowledge. That concept is best dealt with now as a synonym under *S. brysiana.*

The Cayman Islands were once dependencies of Jamaica. Geologically speaking, however, they were once attached to Cuba. Fawcett, a fellow of the Linnean Society and director of Public Gardens and Plantations in Jamaica, sent plants to Kew in 1888. They flowered in June 1901.

Fluteplayer's Schomburgkia

Belize, Guatemala, Honduras, México, Nicaragua?

(PLATES 31, 32)

Schomburgkia tibicinis Bateman. 1841. *Orchidaceae of Mexico and Guatemala*, t. 30, and *ex* Lindley *Bot. Reg.* 27, misc. 57.
Subgenus *Chaunoschomburgkia*

Synonyms

Epidendrum tibicinis (Bateman) Bateman *ex* Lindley. 1838. *Bot. Reg.* 14:8.
Schomburgkia tibicinis var. *grandiflora* Lindley. 1845. *Bot. Reg.* 31, t. 30.
Schomburgkia grandiflora Hort.
Cattleya tibicinis (Bateman) Beer. 1854. *Prakt. Stud. Fam. Orch.*, p. 215.
Bletia tibicinis (Bateman) Reichenbach f. 1862. *Walp. Ann. Bot.* 6:429.
**Myrmecophila tibicinis* (Bateman) Rolfe. 1917. *Orchid Rev.* 25:51.
Laelia tibicinis (Bateman) Williams. 1941. *Darwiniana* 5:77.

The name of this orchid derives from the Latin *tibicen* meaning "flute player." A *tibia*, or flute, in its simplest form is a hollow reed with holes cut at the proper distances. The same Latin word is the basis for the name of one of our leg bones, the tibia, and it can also refer to a pillar, presumably a fluted one. The allusion in this species is to the hollow flute-like pseudobulb with a cleft or opening at the base. Though there have been no accounts of anyone but children playing tunes on this flute, the hollow does provide a home to biting black ants, as anyone who has collected plants of this species can testify. The plant is called *caño* (pipe or tube) in vernacular Spanish, and it was discovered by George Skinner in Honduras. The species has also been called the "horned orchid" in Belize, an allusion to the two divergent lobes of the anther cap. The variety *grandiflora* seems to have been an unusually fine clone but is not truly a variety by modern criteria.

The many synonyms show that this species has been systematically placed in various genera of this alliance, reflecting changing opinion about the genus *Schomburgkia*. When first described by Bateman in 1838, only the vegetative parts of the plants were known, and Lindley wrote that Bateman said the flowers were "the size and color of *Cattleya labiata*." This was apparently wishful thinking, as the flowers are about 8 cm in the best of circumstances and are normally 5 cm in their spread.

Bateman was no doubt misled by the size of the plant and the typically long flower stalk, which can in nature be 1 cm in diameter and as long as 3 m; it is usually shorter under cultivated conditions. It is also possible that he was confusing it with *Schomburgkia superbiens* that does have the large cattleya-sized flowers. In any case, he first had the plants in his collection at Knypersley in 1836 but did not flower them for 4 years. The plants are tough and hard, having 2–4 oval, stiff leaves topping the conical pseudobulb that shows clearly 4–6 nodes by the encircling scars. The flower stalks can produce up to 30–40 flowers in succession. Hamilton's data show peak flowering in May.

Both the sepals and petals are undulate, the sepals being wider than the petals. They are colored magenta or purple, though sometimes tinged with bronze or orange. Usually paler on the reverse in variable white or purple, the central region of the sepals may be almost white. The ovary and pedicel are purple to match and the floral bracts are so small that they ordinarily go unnoticed. The lip is white with much magenta veining in the throat, bright yellow on the callus and sometimes on the rolled back margins of the lateral lobes, and the midlobe is tipped with purple. There are 5–7 distinct keels on the disc. The wide-veined throat and the large curled-up lateral lobes of the lip dominate the appearance of the flower, the midlobe being short and small with a notched apex and lacking an elongated isthmus or base. The anther cap is white, and the column is shaded white and purple and protrudes expectantly within the open space provided by the lateral lobes. The column is somewhat curved with winged edges toward the apex.

There are certain color variants that tend to brown-yellow with paler yellow or pinkish colors on the reverse of the sepals. These may be natural hybrids with *Schomburgkia brysiana.* If so, the name for these plants would be *S. rechingerana.* (See the discussion under *S. brysiana* for more details.) These two species have been often mixed up or considered together as *S. tibicinis,* so that caution is necessary in making identifications and therefore in determining actual distributions.

Undulate Schomburgkia

Colombia, Costa Rica, Trinidad, Venezuela

(FRONTISPIECE; PLATES 33, 34)

Schomburgkia undulata Lindley. 1844. *Bot. Reg.* 30: sub t. 23, misc. 13. 21.
Subgenus *Schomburgkia*

Synonyms

Bletia excelsa Bradford. 1846. Herbarium name. Bradford manuscript on Trinidad at Kew.
Schomburgkia violacea Paxton. 1849. *Paxt. Mag. Bot.* 16:123.
Cattleya undulata (Lindley) Beer. 1854. *Prakt. Stud. Fam. Orch.,* p. 215.
Schomburgkia quesneliana Hort. Paris. 1860. *Pescatorea* 1, fig. 32.
Bletia undulata (Lindley) Reichenbach f. 1861. *Walp. Ann. Bot.* 6:420.
Schomburgkia lueddemannii Prilleux. 1862. *Jour. Soc. Imp. Hort.* 8:123.
Laelia undulata (Lindley) Williams. 1941. *Darwiniana* 5:76.
(Note: The description under *Schomburgkia lueddemannii* in Curtis's Botanical Magazine 138, t. 8427, 1912, is correct for that species, but the plate is *S. undulata.*)

Schomburgkia undulata was first described as coming from the neighborhood of Bogotá at about 700 m elevation. Wagener, the collector, noted that it was an imposing and beautiful plant growing on ruins and cultivated on the tops of walls. Its hallmark characteristic at the time was the elongated and narrow, strongly undulate or wavy sepals and petals with no marginal or other yellow coloring on them. The sepals and petals are rich brown-purple, and there is a violet-red or purple lip with three main keels and an oval, acute midlobe. The lateral lobes are oblong and erect around the column. The disc is white and convex where the keels are especially noticed.

A plant flowered for the first time in 1844 in the collection of Rucker in Wandsworth, England, and Mrs. Lawrence also had some fine specimens. Linden had discovered the plants in 1841 growing on rocks near Truxillo, Venezuela, and also around Pandi near the natural bridge of Icononzo, in Bogotá Province of Colombia, at 800 m. We are told that the Truxillenos called the plant *mulattita* because the curly flowers reminded them of certain women's hair. The plants are well illustrated in Dunsterville and Garay's *Venezuelan Orchids* (1959–1976, 2:256), with notes telling that the species grows at elevations of 30–150 m, sometimes as high as 975 m, in open, light forest near the coast—a semixerophytic area with a strong dry season. Hamilton's compilations show a peak of flowering in February and March in cultivation.

In specimens I have measured the dorsal sepal is 9–13 × 33–43 mm; the petals are 7–11 × 28–39 mm; the flattened lip 17–20 wide × 25–27 mm long. Bracts are 7–10 × 55–60 mm and the pedicellate ovary is 80–90 mm. The three central keels are most prominent with wavy crests both on the disc and near the base of the lip where there is a hollow, raised area. This structure is analagous to the one on the *Schomburgkia crispa* lip and must help in guiding the pollinator proboscis to the nectary as well as effecting pollen removal as the proboscis is withdrawn. The two lateral veins adjacent to the three central veins are also enlarged, though they may not be crested. The pedicels, ovaries, bracts, and columns are a deep to pale rose-pink and there is the hallmark white spot on the disc of the lip.

Plants display typical schomburgkia qualities, with pseudobulbs to 25 cm that have a close-fitting sheath of bracts, and usually two leaves to 30 × 5 cm. As is typical of other schomburgkias in this subgenus, the pseudobulbs are distinctly fusiform—that is, they are thickest near the middle and taper, spindle-like, toward both ends. In this species the lower taper is more elongated and gradual than the upper, with a less pronounced decrease in diameter. The brown flower stalks may reach 1.5 m and bear up to 20 flowers concentrated near the end. The floral bracts are not as long as the ovaries and curl into tubes, hanging down along the main stalk as the blossoms open.

Schweinfurth (1967) has reported this species among the orchid collections in the Guayana Highland of Venezuela, where they had rich pink lips shaded toward white at the apices of the midlobes. The plant collections reported from Trinidad may reflect cultivated plants, but that cannot be ascertained at present. Geologically, Trinidad is now an island that was once part of an ancient land and mountain extension from Venezuela to the east. It has many of the same orchid species that are found on the continent proper.

Kraenzlin described *Schomburgkia campecheana* from the Gulf of Campeche, México, as a natural hybrid between *S. undulata* and *S. thomsoniana,* but as we understand the distribution of the species today and their characteristics, it better fits as a synonym under *S. brysiana* (please see). In Schlechter's *Die Orchideen* Brieger, Maatsch, and Senghas (1976–1981) consider *S. wallisii, S. weberbaueriana,* and *S. elata* to be synonyms of this species, but I am presenting them separately in this book. Other authors have from time to time confused *S. undulata* with *S. crispa,* but that has not been a particular problem.

Schomburgkia heidii, a natural hybrid from Venezuela, results from crosses between *S. rosea* and *S. undulata.* It is discussed under *S. rosea.* *Schomburgkia quesneliana* is a synonym of *S. undulata* and should not be used with *S. superbiens* as was published by Kennedy (1979).

Weberbauer's Schomburgkia

Peru

(PLATES 35, 36)

Schomburgkia weberbaueriana Kraenzlin. 1906. *Englr. Jahrb.* 37:527.
Subgenus *Schomburgkia*

Synonym

Laelia weberbaueriana (Kraenzlin) Schweinfurth. 1944. *Bot. Mus. Leaf., Harvard Univ.* 11:180.

Schomburgkia weberbaueriana has been listed as coming from Colombia and Venezuela, as well as from Peru. But, the type specimen collected at Santa Ana near Cuzco by Weberbauer, a German botanist, was unquestionably Peruvian, and one may well question the identities of the specimens that would vouch for the other two countries. The flowers of this species look much like those of *S. marginata* and may be identical. As I stated in the discussion of that species (please see), I am keeping the two as separate entities in this work because of their geographic disjunction. The same would apply to the relationship of this species with *S. wallisii,* whatever it may be. More field research would resolve the problem if specimens could be found in the intervening areas. The plate labeled *S. weberbaueriana* in Dunsterville and Garay (Vol. I 1959–1976) is now thought to be *S. crispa.* Even the experts are sometimes stumped by this genus.

The original plants of *Schomburgkia* were found growing on boulders in a stream at 1300 m. In *Orchids of Peru* (1959), Schweinfurth gives a good description prepared from a photograph of the type plus the original description. The plants have the conventional fusiform pseudobulbs to 37 cm high. With the flower stalk, the plants reach over a meter. Both pseudobulbs and flower stalk are, as usual, completely enclosed in tight-fitting, papery sheaths. The papery floral bracts are nearly equal in length to the pedicellate ovary and may be 7 cm long.

The sepals and petals are yellow-brown tesselated with darker veins. The sepals are abruptly pointed with wavy margins and measure about 8–10 × 28–30 mm. The petals are similar, but they are a little shorter, and the tips are more rounded. They have yellow margins, but the sepals may or may not have this color marking, which may also be observed in flowers of *Schomburgkia marginata.* The lip is white, arcuate, and has upturned sides. It is about 2 cm long, 1.3 cm wide at the base, broadly oval in overall shape when flattened, and has poorly defined lateral lobes. Kraenzlin said the lobes were best noted by the veining pattern rather than by the lobing itself. The tip of the lip is slightly indented and may have a small point. There are five raised veins passing over the disc, slightly keeled toward the apex; the central and lateral veins are highest. The column is white and 1 cm long.

Jones (1969) made this species a synonym of *Schomburgkia marginata;* Schlechter (1913) made it a synonym of *S. wallisii* but with a question mark. In the text, Schlechter said that he could not distinguish between the two and that Kraenzlin was correct in saying it was close to *S. marginata*. At that time, Schlechter apparently made a distinction between *S. marginata* and *S. wallisii*. In his discussion he further mentions that the lateral lobes of the lip are strongly truncated and therefore do not project forward very much; also, the flowers are yellow-brown with a white lip.

The apparently dark brown flowers of *Schomburgkia wallisii,* with the yellow-white area on the disc of the magenta-edged lip and magenta column, should make this species readily distinguishable from the yellow-brown flowers of *S. weberbaueriana* with white lip and column. This mention of *S. wallisii* characteristics, incidentally, is about the best we have to go on for that species. *Schlechter III* considers *S. wallisii* as well as *S. weberbaueriana* to be synonymous with *S. undulata.* Distinctions from *S. moyobambae* are discussed under that species. This disinformation will have to be ignored at present.

Wendland's Schomburgkia

Guatemala, Honduras, Nicaragua, Costa Rica?

(PLATES 37, 38)

Schomburgkia wendlandii (Reichenbach f.) Jones. 1961. *Am. Orch. Soc. Bull.* 30:638.

Subgenus *Chaunoschomburgkia*

Synonyms

Bletia wendlandii Reichenbach f. 1861. *Walp. Ann. Bot.* 6:431.
Laelia wendlandii Reichenbach f. 1861. *Walp. Ann. Bot.* 6:431.
**Myrmecophila wendlandii* (Reichenbach f.) Kennedy. 1979. *Orchid Digest* 43(6):204–211.

The type specimen of *Schomburgkia wendlandii* was collected in Guatemala by Wendland. The plants are often found on mossy or lichen-covered rocks in full sun. They also can be epiphytic on trees in slightly more moist situations in semiarid countryside. Although it is not a spectacular plant, it hardly deserves Kennedy's comment. "Anyone unfortunate enough to acquire this plant in a shipment of Guatemalan orchids should consign it instantly to the compost heap!" It is an unprepossessing species, and though its flowers are the smallest in the genus, they nevertheless have an interesting characteristic.

The flowers of this species share with *Schomburgkia lyonsii* the quality of non-resupination. This probably indicates something about the nature of the pollinators, but I have not discovered any observations on this point. The sepals and petals of the flowers are dull brown-green and the lip is a dull cream on the midlobe and more green on the lateral lobes. We do not know whether they are non-resupinate because they have not revolved

180° on their axes, or because they have made a complete 360° turn. The description of the plants in Ames and Correll's *Orchids of Guatemala,* for instance, never mentions resupination, nor does Kennedy, Jones, or Reichenbach. As many as 25 flowers may be produced on a paniculate inflorescence.

Hamilton (1986) has eight flowering records, peaking during March. Jones (1968a) describes the plants as being common in lowland Nicaragua. The flowers are sometimes cleistogamous.

DOUBTFUL SPECIES

Schlechter's Schomburgkia

Venezuela

(PLATE 39)

Schomburgkia schlechterana Jones. 1967a. *Fedde Repert. Spec. Nov.* 76(3):193–199.
Subgenus *Schomburgkia*

I have pondered long over what to do about this taxon, whether it represents a separate species population, a color variety of a previously named species, or something else. Unfortunately, information about the plants is fragmentary or nonexistent; only the type description is available. The herbarium record is so poor as to be nonexistent. Jones described the taxon during his studies of the genus, naming the species after Rudolf Schlechter (1872–1925), who had published the first monograph of *Schomburgkia* in 1913 and who later added four new species to the group. According to Jones's account, the sepals and petals are brown-yellow to orange; the lip is white shaded with rose and has three keels. The sepals are narrowly linear, obtuse, 4 × 30 mm; the petals are a little shorter; the lip is 20 mm long and 15 mm wide.

As a result of an unnamed "correspondent in Venezuela" who sent him flowers (no date), Jones was convinced they represented a new concept. He further stated that the plants were cultivated in Barbados and Trinidad, though why he hadn't noticed them previously in preparing his complex of obscure articles on *Schomburgkia* is not mentioned. The type specimen consists only of the Venezuelan flowers (faded or aged for color?) sent by the unnamed correspondent and filed in Jones's private herbarium under #91; the native locality of the plants in Venezuela was never revealed. So there is an element of ambiguity and mystery or question about the whole situation. No drawings or photo illustrations of the new species enlivened Jones's articles or have ever been published elsewhere.

Two plants that I observed in the Caracas Orchid Show in 1982, however, persuaded me that Jones was possibly correct in describing this new species. I had taken notes, photos, and a flower specimen each after the show. The plants were not named, and people at the time were considering them as albino forms of *Schomburgkia undulata* because of their lack of anthocyanins and their pale green-old gold colors. One form, 'Verde', had been granted a JC/AOS at the show as an unusual, pale green color form of *S. undulata*. The other clone, 'Elisabeth', was given a CHM/AOS because of its rare yellow-green color and good form. These awards were apparently never finalized. They do not appear in our published American Orchid Society judging records, for whatever reasons, including a lack of confirming identification. Such forms are not described in Dunsterville and Garay's *Venezuelan Orchids,* the standard reference source for that country, nor are they mentioned in Schultes's *Orchids of Trinidad and Tobago* nor Foldats's orchid accounts in *Flora de Venezuela*.

No herbarium material is available for study, and there has never been an illustration. You can therefore understand why I have reservations and certain skepticism about the species *Schomburgkia schlechterana,* and whether the plants I saw could even qualify for that identification. The orchidists of Caracas now think that they are definitely examples of *S. undulata*. Jones (1970a) published "A Note on the *Schomburgkia undulata* Complex" in the *Orchid Review,* reiterating that he was still convinced that his distinctions among *S. undulata, S. lueddemanii,* and the new *S. schlechterana* were valid. If these two specimens I describe here are not this species, their identity should revert to *S. undulata* for lack of an alternative.

Since writing the above remarks I have been in communication with Dr. Germán Carnevali at the Missouri Botanical Garden. Dr. Carnevali is a Venezuelan botanist and orchidologist and knows more about these schomburgkia plants in question. He writes that there are several albino clones, such as those I am describing, in Venezuela, and that they are considered there to be color forms of *S. undulata*. In that case, *S. schlechterana* would merely become a synonym under *S. undulata* or be relegated to varietal or clonal status. I do appreciate the insight Dr. Carnevali has given me on the problem and hope that further field work and more observations will resolve the placement of these plants.

There is another point to make here. Few if any color variations are reported previously in this genus except for a few clones of white or albescent forms that have turned up among plants of three species: *Schomburgkia humboldtii, S. lyonsii,* and *S. superbiens*. Other color variations, beyond paler or darker clones, seem to be rare or nonexistent. Why shouldn't other color variants appear in schomburgkia populations as they do in cattleya populations? Forms with carotenoids, but lacking anthocyanins, are termed xanthotics. Now, what would these forms lacking both anthocyanins and carotenoids be called? Chlorotics? I doubt it!

My specimens show the sepals and petals were an old green-gold color, and the lip was white with five prominent keels. In my example of 'Verde', the dorsal sepal measures 8 × 32 mm, the petals 7 × 28 mm and the lateral sepals 8 × 30 mm. The lip flattened was 17 mm across the lateral lobes and 26 mm long. The white pedicel and ovary were 6.8 cm long and the white column, strongly curved, about 18 mm long. The floral bracts were about half the ovary length, tightly quilled and papery as they dried. The plant itself was a typical medium-sized schomburgkia. More information on this concept will be welcomed. In the meantime, there is no way to determine the nature of *Schomburgkia schlechterana* and what exactly was in Jones's mind when he first described it.

Wallis's Schomburgkia

Colombia

(PLATE 39A)

Schomburgkia wallisii Reichenbach f. 1877. *Linnaea, Lit Bericht* 41:115–116.

Subgenus *Schomburgkia*

Synonyms

Bletia wallisii Reichenbach f. 1877. *Linnaea, Lit. Bericht* 41:115–116.
Schomburgkia crispa var. *wallisii* (Reichenbach f.) Jones. 1968d. *Israel Jour. Bot.* 17:196–199.

Schlechter claimed in his monograph (1913) that he could not distinguish between *Schomburgkia wallisii* and *S. weberbaueriana,* so *S. weberbaueriana* was relegated to synonymy. The original plant of *S. wallisii* came from Colombia, where it grew at 1300 m and was collected by Gustav Wallis. It was described in 1877 by Reichenbach. We have here another set of confusions so common in this genus. The *S. weberbaueriana* is a Peruvian species well isolated in its distribution from *S. wallisii.* It would seem to me now, after study of available information, that the two are separate species. We never have enough herbarium specimens or field observations of certain species to solve these controversies completely, and there is no record of a type specimen for *S. wallisii* ever having been prepared. In addition, the type description is minimal with no sizes given and no comments or discussion. All we learn is that the lip is three-lobed (what else?!), the midlobe is oblong and acute and the lateral lobes are obtusely angled. Very frustrating! In comparing this species with *S. weberbaueriana,* Schlechter states that it had red-brown flowers with a yellow-white area on the disc of the magenta-edged lip, and that the column was magenta. Perhaps he actually saw Reichenbach's original specimen of *S. wallisii* to obtain these descriptive points, but he makes no mention as to his source of information, only that he is confused.

There is nothing on the history of the species nor its introduction into cultivation except, to quote Reichenbach, "*Pro hortis* . . . N. Grenada." Rolfe (1913) wrote about *Schomburgkia wallisii* and felt that "it bears a very general resemblance to the Venezuelan S. Lueddemannii." He did not say, however, that he thought the two were the same species, only that it was interesting to recognize the plant after such a long interval. There is no mention of where the plant originated, only that it was flowering at Kew. The flowers were apparently dried after the blooming; the herbaria of Kew and the New York Botanical Garden each have a specimen made from it.

The labels say only *ex hort* Kew with no date and add that the plant came to Kew from Glasnevin Botanical Garden. I find at the New York Botanical Garden that the specimen shows flowers of *S. splendida,* so we are still no further ahead on *S. wallisii.*

After studying the available literature, I feel that *Schomburgkia wallisii* may be the South American counterpart of *S. lueddemanii,* the latter seemingly centered in Panama and Costa Rica, while the former is found in Colombia and Venezuela—a disjunct relationship much like that of *Cattleya aurea* and *C. dowiana.* As stated before, some new field work might resolve these suppositions. Otherwise *Schomburgkia wallisii* should be forgotten as a *nomen obscurum* or placed in synonymy with *S. lueddemanii* as Rolfe suggested.

One more point can be mentioned. Some of the South American flowers of *Schomburgkia lueddemanii* seem a darker brown with brighter magenta than the Panamanian-Costa Rican types, which are a more definite tan. This is particularly evident in the border about the lips, as mentioned by Schlechter. The South American flowers are also a little larger than the forms from Central America. In Kennedy's *Orchid Digest* article (1979), a flower with the redder color pattern is labeled *S. lueddemannii.* It is the only form I know that Reichenbach could have identified as *S. wallisii.* This color variation is otherwise the only distinctive schomburgkia form I have been able to turn up that does not already have a distinguishing name.

I have just seen a plant of this *Schomburgkia wallisii* type from the Smithsonian Institution Collection, #781332, labeled *S. leuddemaniana.* The stalks produce 6–8 flowers each with hanging, ruffled sepals and petals of dark chestnut brown. The column and reverse side of the lip are intense magenta that also margins the lip and halos the yellow disc at the base of the midlobe. The anther cap is also yellow. The three enlarged central veins are white in the throat and yellow on the disc. The pedicels are greenish with a slight purple flush, and the brown, scarious flower bracts are about half their length. The measurements are as follows: dorsal sepal 8 × 54 mm; petals 9 × 43 mm; lateral sepal 7 × 45; lip flattened 14 × 20 mm across lateral lobes, midlobe 10 × 13; pedicel and ovary 100 mm; bract 60 mm.

I will press a stalk of this plant for deposit at the Ames herbarium as a neotype of *Schomburgkia wallisii* as it is understood today.

CHAPTER 2

The Genus *Sophronitis*

HISTORICAL AND CULTURAL INFORMATION

The genus name *Sophronitis* is derived from the Greek *sophron,* meaning chaste, modest, and, in this case, small. The suffix *-itis* indicates a close connection, referring this genus back to *Sophronia,* a genus of small orchids with no botanical relationship to *Sophronitis*. For the most part, it is hard to imagine a group of orchids with flowers less modest or less chaste—the brilliant colors of some clones are flashy and outstanding and always make them instant favorites. These colorful species are the choice parents for breeding red or orange flowers on the miniature plants that are so popular today. The enthusiasm for this group is easy to understand.

The genus was founded by Lindley in 1828 when he described *Sophronitis cernua,* indeed a comparatively modest species next to its flamboyant relative, *S. coccinea*. The essential characteristics of the genus are the small plants with clustered pseudobulbs; the lip sessile on the base of the column; small wings on each side of the stigmatic cavity; a short, somewhat thickened column; and eight pollinia. We also know that most of the flowers in this genus are colored some shade of red, though variants may be yellow, salmon, coral, vermilion, rose, or magenta-toned. Only one or two white forms have ever been mentioned, and even the yellow forms are decidedly rare and choice horticultural prizes. The Brazilian local name for a sophronitis, incidentally, is a *vermelho,* referring to its color.

The various species are all mountain dwellers and drew immediate attention to the *Orgãos* (Organ) Mountains north of Rio de Janeiro. Subsequently, other ranges, such as the Serra do Mar and the more inland Serra da Mantiqueira, both paralleling the coast between Rio de Janeiro and São Paulo, were found to be rich habitat areas. Dr. Fowlie has climbed the ridges and written about finding some of these plants in various locales, among them the Pico de Açu in the Organ Mountains above Correos, reaching 2240 m elevation, and the Serra do Espinhaço. The newly described species by Francisco Miranda, *Sophronitis bicolor,* also comes from these coastal mountain areas from Espírito Santo to Rio Grande do Sul.

These mountain environments should be reflected in *Sophronitis* culture that requires cool, humid conditions with good air and ample drainage. The plants require less light than cattleyas. These cultural parameters are passed along to most of their hybrids—particularly those of miniature stature. Dressler (1981) suggests that the color of their flowers and the structure of lip and column would indicate that these flowers are pollinated by hummingbirds, but no one has made a study of this facet of their existence—no one but Dressler even mentions it.

The plants of most species are famous for their temperamental flower production. They must be well established and vigorously growing to produce fine flowers. If numerous large flowers are produced one season, you can pretty well count on fewer and smaller flowers the next. In fact, you might not even recognize that the flowers were produced by the same plant! A sustained excellence of culture is paramount for these plants to reach their potential. With good conditions over a period of a few years, specimen plants can be produced in small pots with several flowers at each blooming. One of the best-flowered clones I've seen was grown on a New York City north-facing windowsill in an old transparent plastic hatbox, so a cool, humid greenhouse is not necessarily a must in providing the ideal conditions. An ideal night temperature is about 50°F.

Flowering occurs as new growths are developing, the unfolding leaf providing protection for the developing bud. A single flower is the rule with all species but *Sophronitis cernua* and *S. pterocarpa.* In fact, be on the lookout for any clones of the other species that might regularly produce more than a single blossom. Some rare clones of *S. coccinea* regularly form two flowers (see Plate 48), but this phenomenon has not been reported for the other taxa. Three flowers at once are unknown at this point.

Humidity is not only required for proper growth but also for full expansion of the flowers from the small buds. The petals, particularly, take a few days to widen fully and will stop enlarging if the humidity is less than perfect. Even so, fold marks appearing near the edges of the petals may never completely disappear, especially on larger flowers, and petals may show slight notches on their upper margins if they were lapped too tightly together as they developed from the tight bud.

The best plants I've seen have been grown in small treefern pots planted in moss-covered bark mixtures, moss itself, or in osmunda fiber. This apparently duplicates, to a large degree, their natural habitat in moss on tree branches in the higher reaches and ravines of the mountains where cloud drifts keep the environment damp. The plants may partly dry out during sunny daylight hours, but dampness is restored with cloud cover in late afternoon. A small pot can provide space for 4 or 5 years of growth when dealing with these small species. The plants should be disturbed as little as possible once they begin to grow well. When repotting, make an effort to disturb the roots as little as possible, and also be conservative about separating the plants into small divisions. They definitely resent repotting and division.

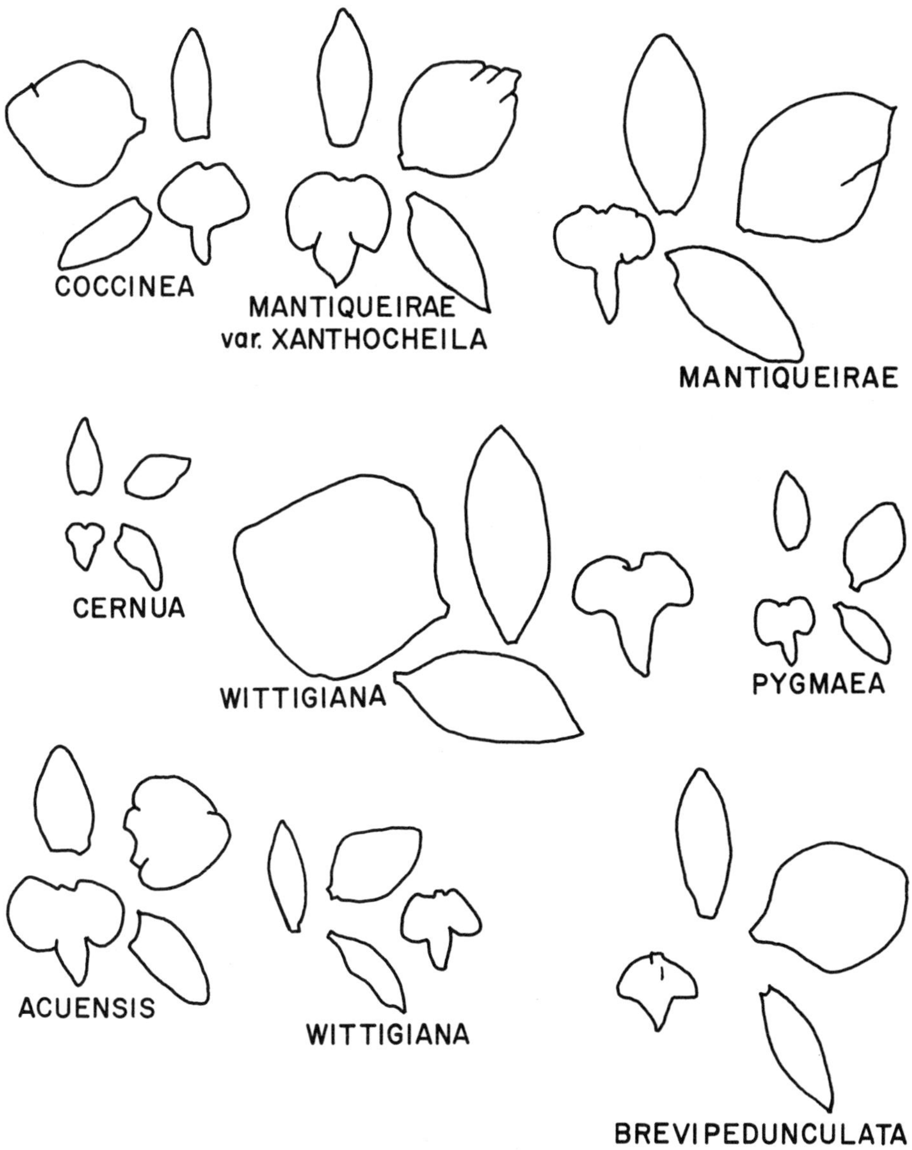

Fig. 2-1. Comparative outline drawings of the flower parts of *Sophronitis* species. Two forms of *S. wittigiana* show the possible variation in size from one flowering to the next.

CLASSIFICATION

The flowers of *Sophronitis* are so often similar to one another that we must turn to habitat location, season of bloom, and the details of vegetative structure to tell one species from another. The nomenclature has been worked out in considerable detail by Dr. Jack Fowlie, who has also named some of the species in this complex. His articles in the *Orchid Digest* (1968, 1972, 1975a, 1975b, 1975c) have been especially helpful, and they should be read by everyone interested in firsthand accounts of habitat information and plant characteristics. (They have been available as a separate reprint from the magazine.)

With only nine species in the genus, a key is not particularly useful or even necessary. In addition, there will be those individuals who may not agree with my treatment of the taxa. Some taxa may be considered species, others only varieties; but after some intense study and observation, this account is my way of evaluating them at present, coordinating past records and terminology with current ideas. A few labels will have to be changed, and I do not know what will be done about registration of some hybrids. Table III lists the present intergeneric hybrids produced from these species.

TABLE III

Intergeneric Hybrid Combinations with *Sophronitis* Registered Through March, 1992, with Approved Abbreviations

SOPHRONITIS	
× *Barkeria* × *Cattleya* × *Laelia*	= *Matsudaara (Msda.)*
× *Barkeria* × *Laelia*	= *Staalara (Staal.)*
× *Brassavola*	= *Brassophronitis (Bnts.)*
× *Brassavola* × *Broughtonia* × *Cattleya* × *Laelia*	= *Hasegawaara (Hasgw.)*
× *Brassavola* × *Cattleya*	= *Rolfeara (Rolf.)*
× *Brassavola* × *Cattleya* × *Epidendrum* × *Laelia*	= *Rothara (Roth.)*
× *Brassavola* × *Cattleya* × *Epidendrum* × *Laelia* × *Schomburgkia*	= *Johnyeeara (Jya.)*
× *Brassavola* × *Cattleya* × *Laelia*	= *Potinara (Pot.)*
× *Brassavola* × *Cattleya* × *Laelia* × *Schomburgkia*	= *Fergusonara (Ferg.)*
× *Brassavola* × *Epidendrum*	= *Epibrassonitis (Epbns.)*
× *Brassavola* × *Laelia*	= *Lowara (Low.)*

TABLE III *continued*

× *Broughtonia* × *Cattleya*	= *Bishopara (Bish.)*
× *Broughtonia* × *Cattleya* × *Epidendrum* × *Laelia*	= *Buiara (Bui.)*
× *Broughtonia* × *Cattleya* × *Laelia*	= *Hawkinsara (Hknsa.)*
× *Broughtonia* × *Laelia*	= *Hartara (Hart.)*
× *Cattleya*	= *Sophrocattleya (Sc.)*
× *Cattleya* × *Diacrium* × *Laelia*	= *Higashiara (Hgsh.)*
× *Cattleya* × *Epidendrum*	= *Stacyara (Stac.)*
× *Cattleya* × *Epidendrum* × *Laelia*	= *Kirchara (Kir.)*
× *Cattleya* × *Epidendrum* × *Laelia* × *Schomburgkia*	= *Izumiara (Izma.)*
× *Cattleya* × *Laelia*	= *Sophrolaeliocattleya (Slc.)*
× *Cattleya* × *Laelia* × *Schomburgkia*	= *Herbertara (Hbtr.)*
× *Constantia*	= *Conphronitis (Conph.)*
× *Diacrium* × *Laelia*	= *Severinara (Sev.)*
× *Epidendrum*	= *Epiphronitis (Ephs.)*
× *Epidendrum* × *Laelia*	= *Stanfieldara (Sfdra.)*
× *Laelia*	= *Sophrolaelia (Sl.)*
× *Laelia* × *Schomburgkia*	= *Deiselara (Dsla.)*
× *Schomburgkia*	= *Schombonitis (Smbts.)*

DESCRIPTION OF THE SPECIES

List of species of *Sophronitis*

S. acuensis
S. bicolor
S. brevipedunculata
S. cernua
S. coccinea
S. mantiqueirae
S. pterocarpa
S. pygmaea
S. wittigiana

Açu Sophronitis

Brazil

(PLATE 40)

Sophronitis acuensis Fowlie. 1975c. *Orchid Digest* 39(3):147–151.

The plants of *Sophronitis acuensis* were first noted in the greenhouses of Orquidário Binot. A local collector had brought them in from the Pico de Açu in the Organ Mountains. The plants were different from the usual *S. coccinea* plants by their "squat and subhemispherical" red-flushed pseudobulbs, shortly-petioled ovate and curved leaves, and flowers on stalks reaching to the top of the leaves. The pseudobulbs had a scattered pattern of formation and were not in an ordered double row as is exhibited by some other species.

Guided by the *caboclo* collector, Fowlie was able to collect plants of this new entity after a climb to 2100 m. At that altitude they grew on small creekside trees in the ravines that extended down the mountain for 200–300 m. Plants in flower were almost in full sun, and those without flowers were in deep shade and not growing so well. The plants blossom mainly from September through November, but there is indication that a May–June flowering is also possible if temperatures are not too cold. The habitat is higher and colder than for other species in the genus.

In the type description, Fowlie notes that the whole plant is only 5 cm high and that one or two flowers are produced. Sepals are 8–12 mm wide × 24–28 mm long; petals are 20–22 mm wide × 25–30 mm long; the lip is 16–20 mm wide × 15–18 mm long with a narrow elliptical apex. The column is 8–10 mm long, and the pedicellate ovary is 25–30 mm long. The reverse of the leaves and the pseudobulbs are often pigmented with maroon.

It is not hard to recognize *Sophronitis acuensis* plants from their vegetative chacteristics, even though the flowers have a typical *Sophronitis* shape. Fowlie describes the color of the flowers in nature as a livid deep red but says they become orange-red once in cultivation. I note a violet or purple cast to the color in cultivated flowers I have seen. There is much yellow with red veining on the narrow lip.

Native temperature and light relationships definitely affect flower color, as we often find out by cultivating orchids in captivity. The pink strains of *Oncidium variegatum, Lemboglossum cervantesii,* or *L. rossii* are other examples of species that change color when brought down from higher altitudes for cultivation. The flowers become a disappointing white, like most of the other clones, until grown on for two or three years. Well-established plants are most likely to produce the original pink colors again.

Fowlie has written me that a clone of *Sophronitis acuensis* found by Arturo Mello above Nova Friburgo had an all-yellow lip. It was consequently called variety *xanthocheila* but should, however, have been given a clonal name instead of a varietal designation. Individual clones of *S. acuensis* are otherwise not yet distinguished in the literature, though one plant has been granted a CBR by the American Orchid Society. There is no picture, however, and there is nothing in the award description that would confirm the identification of the awarded plant. On the contrary, the described rosy salmon color could indicate that it is *S. brevipedunculata*.

Some clones of *Sophronitis cernua* have recently been sold under the label *S. acuensis*. Even if you pronounce the name correctly, "ah-soo-en-sis," the plants are still mislabeled and should be changed to *S. cernua* unless, of course, they are single-flowered and fit the description above. In Latin, incidentally, accents or other diacritical marks are not included, so the epithet is written with only a *c*. The deep red color of the flowers should make it a good species for hybridizing.

Bicolored Sophronitis

Brazil

(FIGURE 2-2; PLATE 41)

Sophronitis bicolor Miranda. 1991. *Die Orchidee* 42(5): 227–230.

This species of *Sophronitis* has been published as the final manuscript of this text is being prepared for the printer. I am pleased to insert the account of *S. bicolor* as I feel it resolves some of the discrepancies observable within this genus. The several varieties that have been allotted to *S. coccinea* are so numerous that the taxon has become a sort of wastebasket species for all that was not known. This new concept removes one group of plants from further consideration under *S. coccinea,* namely the larger sized plants called var. *tetraploidea* or *gigantea* that were presumably tetraploid clones. It does not mean that there are no tetraploid clones of *S. coccinea,* but it does mean that the evidence for their tetraploidy as well as their identification will have to be reevaluated.

Miranda of Orquidário Boa Vista in Rio de Janeiro observes in his description that the plants of *Sophronitis bicolor* can be recognized even out of flower because of their elongated leaves reaching 15 cm—putting the foliage on a par, he states, with the leaves of *Laelia pumila*. The plants may also be recognized by their loose-tipped bracts, sheathing the pseudobulbs on newly developing growths, and the 5 cm flower stalks that are significantly longer than those of other species. The tip of the flaring top bract encloses the leaf base and newly developing flower bud.

In addition, the flowers show broader lips than those of other species, and the lips also show more yellow with the red veining. This color characteristic may cause them to overlap with the variety *xanthochila* of *Sophronitis coccinea,* as well as the variety *gigantea* mentioned above. Perhaps both of these varietal categories should be transferred to this species altogether when more is known about it.

Miranda notes that the plants may naturally have red pigmentation that is not related to light exposure. The smooth, cylindrical pseudobulbs may be reddened as well as the leaves. I've seen that the leaves on most plants have the same red midribs and margins as do *Sophronitis coccinea* leaves, though they are longer and more erect with petioled bases. The pseudobulbs are 1–1.5 cm apart on the rhizomes. The sepals and petals vary somewhat in color from more orange tones to redder colors. The anther cap is purple.

We anticipate more information on this species as it becomes more

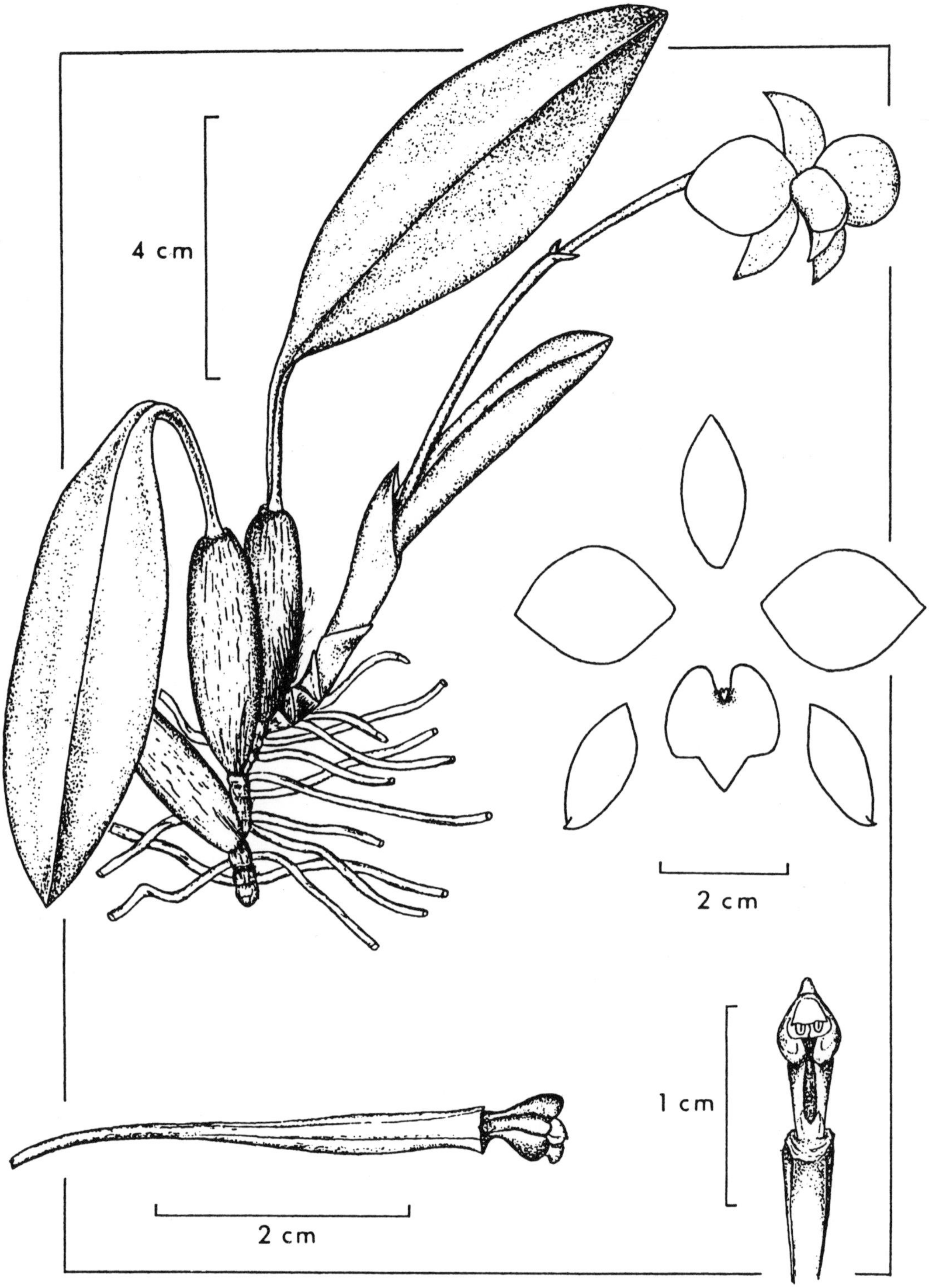

Fig. 2-2. *Sophronitis bicolor*. Diagram of plant habit and flower detail. Used with permission of Francisco Miranda and courtesy of *Die Orchidee,* published by the Deutsche Orchideen-Gesellschaft E.V.

generally recognized. Plants are already in cultivation, though they are probably labeled *Sophronitis coccinea* instead of *S. bicolor*. Its further use in hybridizing is inevitable, and we hope the correct name will now be used.

Short-stalked Sophronitis

Brazil

(PLATES 42, 43)

Sophronitis brevipedunculata (Cogniaux) Fowlie. 1972. *Orchid Digest* 36(5):181–194.

Synonym

Sophronitis wittigiana var. *brevipedunculata* Cogniaux 1898–1902. Martius' *Flora Brasiliensis*, vol. 3, part 5, p. 318.

Sophronitis brevipedunculata is another sophronitis species that can be identified by a close look at its vegetative habit of growth. It typically produces a regular double row of pseudobulbs with ovate green leaves flattened to parallel the surface of the trunk or branch upon which the plant grows. This type of growth habit is seen only in two other forms of sophronitis—*S. cernua* var. *mineira* and *S. wittigiana*. The growth habit is well pictured in Fowlie's 1972 paper. He writes that the flowers tend to have a "salmony tone of reddish coloration quite unlike colors evidenced in *S. coccinea*." The leaves of this species are more elongated than either of the above-mentioned forms. The flower stalks are ordinarily very short, the characteristic that gives the species its name, so the ovary of the flower appears to emerge almost directly from the top of the pseudobulb. With less light this stalk can enlarge to some extent, but it is still comparatively short for the genus.

Another characteristic of this species is that the plants usually grow only on *Vellozia* bushes—an unusual host plant that grows in drier, cool mountain habitats, sometimes forming whole copses at the 1500–2000 m level. Some of the orchids may also grow directly upon the rocks. The *Sophronitis brevipedunculata* plants are collected simply by cutting off the branch upon which the plant is growing, and the plant may still be on that branch when bought for cultivation. The *Vellozia* scrub grows along with cacti and other thorny, dryland vegetation. Accordingly, these plants of sophronitis can take more light and less water than plants of *S. coccinea*, but they require the same sort of coolness. In nature they exist mostly on condensation from air currents and the cloud caps that form over the high mountains during the late afternoon and night, not from seasonal rains.

For those with greater botanical curiosity, *Vellozia* is a genus of shrubby, sometimes treelike plants with fibrous woody stems covered in a thatch of old leaf bases (Smith, 1962a). The stems are dichotomously branched. These unique plants are botanically placed in their own family, which is closely related to the amaryllis or lily family. There are about 65 species in tropical and southern Africa, Madagascar, and Brazil, a distribution that indicates great age for the group, back to the time when all these areas were part of Gondwanaland, before the continents were separated by continental drift. These bushes also provide the habitat for plants of the genus *Pseudolaelia*. A *Vellozia* and *Sophronitis brevipedunculata* are both well illustrated in photo 28 in *Native Orchids of Brazil* (Assoc. Orquid. de São Paulo, 1977).

The flowers of *Sophronitis brevipedunculata* are slightly different in that the petals are "abruptly pointed" at their tips, and they have a different sort of red color, as already discussed. The type description shows the flower measurements to be as follows: sepals 5–8 mm wide × 18–20 mm long; petals 20–25 mm wide × 20–35 mm long; lip 18–20 mm wide × 26–30 mm long with narrow elliptic shape and streaked with some yellow veining. The column measures 8–12 mm, and the ovary with stalk 25–35 mm. As usual, there is a lot of variability in the flower size, depending upon the clone and the culture. These plants grow in southern Minas Gerais, extending westward to the Serra do Espinhaço and then eastward to Caparão. The blooms are formed in nature during April and May, and in cultivation from October through December.

Fowlie mentions a "blue" variety turning up in a shipment of otherwise normally colored forms. Other color variants are not described at present, though a yellow clone is illustrated in *Native Orchids of Brazil*. One unusual color variant with superior flowers that should have been identified under this species was awarded as *Sophronitis rossiteriana* 'Ericka' at the Eastern Orchid Congress in 1967, which was held at Longwood Gardens, Pennsylvania, and hosted by the DuPonts (Plate 43). The plant is long gone and the award was never made final. There was valid questioning at the time and much discussion about a correct name, as this species had not yet been recognized or described. The orchid belonged to Margaret Ilgenfritz of Ilgenfritz Orchids and created a sensation at the time among those who saw it.

But *Sophronitis rossiteriana* flowers were originally described as having petals of sulphur yellow with obscure orange-veins and upright pseudobulbs with leaves slightly convex, intensely green, and shining. The plant first came from Minas Gerais near Barbacena and was collected, flowering in June, by P. Binot. (See accounts of *S. coccinea* and *S. mantiqueirae* for more details, along with Plates 49, 51.) *Sophronitis rossiteriana* is separated in Cogniaux's key in Martius from *S. coccinea* by having more or less erect, acute leaves, a different "orangey" flower color, and pointy petals twice as wide as the sepals. The leaves are described as being erect, outspread, slightly arcuate, 4–9 × 10–25 mm. The key separates these two from *S. wittigiana* on the basis of leaf shape.

Keying this Ilgenfritz plant in Martius, it comes under "leaves broadly oval, narrowly pointed, peduncle longer than leaves," while *Sophronitis rossiteriana* falls under "leaves oblong or ovate-oblong, obtuse to slightly pointed, peduncles shorter than leaves or only slightly longer." The sum of all this is that the 'Ericka' plant did not properly match any of the previous species concepts. In fact, it could not have been properly identified in 1967. It does match *S. brevipedunculata,* now that *S. brevipedunculata* has been raised from varietal status and recognized as a separate species through the work of Fowlie. *Sophronitis rossiteriana* is considered here now as a color form under *S. coccinea,* though I am not positive it should not be placed under the concept *S. mantiqueirae*. One quickly finds that our knowledge and definition of sophronitis species is still spotty.

Nodding Sophronitis

Brazil, Paraguay

(PLATES 44, 45)

Sophronitis cernua Lindley. 1827. *Bot. Reg.* 13, t. 1129.

Synonyms

Sophronia cernua Lindley. 1827. *Bot. Reg.* 13, t. 1129.
Sophronitis modesta Lindley. 1828. *Bot. Reg.* 14, sub. t. 1147.
Sophronitis hoffmannseggii Reichenbach f. 1842. *Linnaea, Lit. Bericht* 16:286. Hoffmansegg., 1840, *icon. ined., Preis. Verz. Orchid.* 27.
Sophronitis isopetala Hoffmansegg. 1842. *Verz. Orch.* 60 in 1843 *Bot. Zeit.* 1:834.
Sophronitis nutans Hoffmansegg. 1842. *Verz. Orch.* 61 in 1842 *Linnaea, Lit. Bericht* 16:286. Hoffmansegg., 1840, *icon. ined., Preis. Verz. Orchid.* 28.
Epidendrum humile Vellozo. 1881. *Flora Flum. Icon.* 9, t. 24; text ed. Netto in *Arch. Mus. Rio,* 1881, 5:36; *non E. humile* Smith.

Recognizing *Sophronitis cernua* has never been difficult nor has the species ever been confused with another. It is the type species for the genus, but the name just would not settle down and stop changing. With its short, almost nodding racemes of five or six small flowers as a hallmark, "drooping" became the species epithet. Another definition of *cernua* in the Latin dictionary is "falling headlong, with the face toward the ground." The word nodding from the synonymous *S. nutans* can still be used for the common name.

The pseudobulbs and leaves grow almost flat against the surface of the trunk or branch. Examples of the species were collected first by William Harrison, a merchant in Liverpool and Brazil, on trees at Botafogo in what is now the city of Rio de Janeiro. He sent the plants to Mrs. Arnold Harrison, a relative, at Aigburgh near Liverpool, who flowered them in her "stove," as tropical greenhouses were called in 1826. One early account tells us that the thriving plants were grown on a branch from an old apple tree, being kept moist and warm in the stove, which mimicked their original seaside habitat.

The pseudobulbs are rounded and somewhat elongated compared to other flat-growing types and grow pretty much in a row. The *mineira* variety (what Fowlie calls a subspecies) from inland Minas Gerais, however, has a double row formation that is distinctive, a characteristic it shares with *Sophronitis wittigiana* and *S. brevipedunculata*. On small plants established in pots, however, instead of upon a slab of treefern or cork, it may be a characteristic that is not so easily observed. The main habitat for the species is the seacoast region north and south of Rio de Janeiro and extending inland to the Organ Mountains and the western slope of the Serra da Mantiqueira and São Paulo State—an extensive area.

The leaves are often a gray-green color, convexly curved, thickly fleshy and leathery, and somewhat folded up along the channeled midrib. The flowers measure about 1.5–2 cm across with petals about 5–6 mm broad and 10–12 mm long. The lip is 6–7 mm wide × 8–10 mm long, rather rigid, pointed, and concave without noticeable side lobes—rather like a pointed scoop. The paler column is always noticed because of its intensely purple wings at the tip. This seems a constant feature and probably has to do with attracting a particular pollinator, perhaps a hummingbird. No one knows.

The flowers can be a cinnabar red with the base of the lip and the column orangey-toned or yellow. The intensity can vary from clone to clone. Variety *lowii* Hort. *ex* Decker was a lemon yellow variety, and other yellow clones have been found as well. One pale yellow one described as an albino by its discoverer, Waldyr Endsfeldz, is called 'Endsfeldzii' and was found in the *mineira* habitat. Two of the yellows without clonal names are pictured in *Native Orchids of Brazil*.

According to Fowlie, the pale, more faded clones coming from near the seacoast are referred to as var. *littoranea*; var. *mineira* is used for inland forms with more carmine-colored flowers. By the current international rules, the *littoranea* name would revert to *Sophronitis cernua* var. *cernua*, as long as other botanical varieties within the species were officially named, in this case the variety *mineira*. Also a variety *albiflora* was named by Cogniaux in 1906, but nothing is known of it today. Technicalities aside, it is difficult or impossible to change local usage of older names except over long periods of time, even though botanical rules may require certain changes. In this genus, especially, there are many such names and at the same time a deficiency of clonal designations that would help separate subspecies, true varieties, and clones or forms from one another.

An inland form of *Sophronitis cernua* collected by Ralph Spencer is pictured in the *Orchid Digest* (1968) and has pale, orangey-colored, almost yellow flowers. The inland plants often have grayer leaves with darker green margins and midrib when compared to more uniformly colored coastal forms. The Hamilton Peak for flowering is from October through December. In nature they flower from May through August. Some plants of this species have been marketed in the United States as *S. acunae*, but the origin of that name and any differences from the usual *S. cernua* are unknown to me. I have heard that that name was intended for coastal types of plants with more gray-green foliage. The two plants I saw, purchased with this label, were *S. cernua*.

Scarlet Sophronitis

Brazil

(PLATES 46, 47, 48, 49, 50, 51)

Sophronitis coccinea (Lindley) Reichenbach f. 1861. *Walp. Ann. Bot.* 6:465.

Synonyms

Cattleya coccinea Lindley. 1836. *Bot. Reg.*, n.s. 9: sub t. 1919.
Sophronitis grandiflora Lindley. 1837. *Sert. Orch.*, t. 5, fig. 2.
Cattleya grandiflora (Lindley) Beer. 1854. *Prakt. Stud. Fam. Orch.*, 209.
Sophronitis militaris Reichenbach f. 1861. *Walp. Ann. Bot.* 6:465.
Sophronitis rossiteriana Barb. Rodrigues. 1877. *Gen. Sp. Orch. Nov.* 1:77.
Sophronitis lowii (Hort.) Curtis. 1923. *Gard. Chron.* 73:40, 44.
Eunannos coccinea (Lindley) Porto and Brade. 1933. *Arq. Inst. Biol. Veg.* 12(2):206, 216.
Sophronitis coccinea f. *rossiteriana* (Barb. Rodrigues) Pabst and Dungs. 1972. *Dic Orchidee* 23(1):20.

Sophronitis coccinea began its botanical existence as one plate of Brazilian orchids in a series of paintings by the French artist, naturalist, and traveller Michel Descourtilz. He called the plant *Epidendrum ponceau* (flame-colored) when he sent the plate (see Plate 47) to Lindley for identification. Thus begins its complicated history of various names, now reduced to synonymy, and described in full detail by Fowlie (1968). For many years we did not properly know what to call the species, and if there is any question at this date about its correct name, please refer to Fowlie's papers.

More has been written about this species and its color forms than about any other sophronitis, and the wealth of information is large, even if the species name may sometimes be incorrect by modern rules. Some of the designations, especially varietal epithets, may be clonal names that are only horticulturally recognized. True varieties, indicating a particular subpopulation of a given species, must be properly published to have an official status. Most are not, and in such cases the abbreviation Hort. may stand after the name instead of an author, such as Lindley or Reichenbach f. Since 1 January 1935, all species and varietal names must be accompanied by a Latin diagnosis to be validly published.

In publishing Descourtilz's plate in *Sertum Orchidaceum,* Lindley (1837–1842) himself said that he had erred in calling the plant a *Cattleya* because he did not have the chance to count the pollinia—eight—until later and that the plant was really a *Sophronitis*. He also admitted that it resembled a small cattleya in its growth, except for its pseudobulb (most cattleyas known to that date were bifoliate and thin stemmed) and the absence of a floral sheath for flower formation. Lindley further wrote that the petals were shorter than the sepals—a condition we know today can result from less-than-ideal flower development.

The identifying characteristics of these *Sophronitis coccinea* plants are their cylindrical rather than spherical pseudobulbs and their upright, rather narrow leaves, which have a red stripe down the midrib and usually a red margin around the edge. Single flowers, rarely two, are produced (see Plate 48), with the Hamilton Peak coming in March for cultivated plants. However, other bloomings may be scattered throughout the year.

Specimen plants may produce as many flowers as the well-flowered clone 'Ethel Sackett'—123 flowers and 14 buds in an eight-inch pot—which was awarded a CCM of 96 points by the American Orchid Society in March 1975. That record has not yet been bested. The flowers of that clone were 5 cm across when awarded, and on other clones, such as 'Big Red', measurements of 7–7.4 cm across were the rule. These flamboyant, larger-flowered types are choice clones that are perfectly cultivated to achieve full flower size. Some are considered to be tetraploids, and might today be reclassified as forms of *Sophronitis bicolor* (please see).

One Japanese orchid grower, Jiro Miura, has made a specialty of using fine clones for both sib crosses and hybrids. The clones in his 'Small World' series, two of which were medal winners at the 11th World Orchid Conference in Miami in 1984, had individual petals measuring 3.5–4.5 cm in length and 3–3.5 cm in width. The petals are nearly round and usually have rounded tips.

The plant in Descourtilz's painting came from what is now called the Serra do Mar, a range paralleling the coast between Rio de Janeiro and São Paulo. Plants of the species have also been found in the Organ Mountains northeast of Rio, and Fowlie also lists the Serra Paranapiacaba, as well as mountains in Espírito Santo, as habitat areas. Several varietal types have been noted by Fowlie and others, mostly describing variations in color rather than in size. Fowlie, in addition, has distinguished three ecotype variations based on different habitats. These ecotypic variants presumably do not change their growth habits when grown side by side under the same conditions in cultivation. Though these names are not validly published to have botanical status, they do help interpret the complicated and variable *Sophronitis coccinea* species population:

> Ecotype *militaris*—has very thin pseudobulbs like "pencil leads," standing close together in rows; plants preferring dark forests where they grow low on the trees and have short stalks so the flowers appear to come directly from the base of the leaf.
>
> Ecotype *coccinea*—has the usual club-shaped miniature cattleya-type of pseudobulb matching the plant in Descourtilz's painting; grows under brighter conditions and has more prominent red-purple midrib and comparatively shorter and broader leaves than *militaris*.
>
> Ecotype *insolaris*—is the rarest type; plants prefer twiggy branches in brightest light; the purple-red on midrib and on leaf margins brightest; leaves lighter green in contrast; pseudobulbs more swollen, but not spherical; flowers considered to be brighter red. This is possibly the group that has produced most of the tetraploid forms.

The color varieties are given below. The quandary over such names has been discussed before: Are they color or pattern categories that include a population of several different clones and should therefore be called a variety and be printed in italics, or does the name indicate a single clone that should be printed in Roman type and enclosed in single quotation marks? I have usually chosen the varietal approach as the descriptions indicate, but it is difficult to tell which is truly correct:

Var. *aurantiaca* Cogn.—this term was used to describe a clone with more orange tones in the flower than usual. Since this kind of change can depend at least partly upon light and temperature fluctuations during growth of the individual flower, the category is probably best left unused.

Var. *barboleta* Hort.—this is likely a clonal variation and should be written 'Barboleta'. Some Brazilians have called it 'Papilio', butterfly. The flowers have the red color concentrated along the veins, with a yellow flush overall to give an orange appearance. The base and center of the petals are bright yellow, matching the base and disc of the red-veined lip. The color pattern is somewhat peloric without much change in petal shape. The petals are a bit longer, narrower and more pointed than those of a normally colored flower. Completely peloric clones have been reported, but I do not know of any illustrations to cite. In Brazil these are also referred to as "labelloid" varieties. They, too, should individually have separate clonal designations.

Var. *carmesa* Hort.—this category of flowers has more purple, or less red, in their color—in Brazilian terminology, *carmesim*. Fowlie says the variation appears once to every 200 "normals" and may be what O'Brien (1886) was referring to in his var. *purpurea*, in which case the *purpurea* name would have priority for this varietal designation. Later, var. *purpurea* was considered to be the same thing as *Sophronitis wittigiana*. Veitch describes *purpurea* (1887, p. 106) as having shorter, thicker pseudobulbs and flowers of a bright carmine-purple—a description that does not sound like typical plants of *S. coccinea*. This variety first appeared in Hofrath Keil's garden at Leipzig about 1850 but was never illustrated. Veitch also tells us that most of the early plants in cultivation were sent by Gardner from the Organ Mountains to Loddiges of Hackney, England, where the first plants had flowered in October 1841. These *purpurea* clones are known to us today only in the literature, and there is still no color illustration on record anywhere to show they are not *S. wittigiana* (please see). I have listed var. *purpurea* in synonymy under *S. wittigiana* but will keep var. *carmesa* here for the time being.

Var. *discolor* Hort.—has flowers that are more of a brick-red with darker pigment along the veins and with a yellow lip. Some know this variety, or closely similar forms, by the name *xanthocheila* Hort., which is the same varietal name used for yellow-lipped forms of *Sophronitis mantiqueirae* or *S. acuensis*—another small source of confusion. Variety *discolor* occurs in specific localities, and Fowlie estimates one in 500 plants for its frequency of appearance. I am also told that whole colonies have been found with this pattern. Individual clones deserve their own cultivar names when shown or awarded, preceded by var. *discolor*. In Latin *discolor* means "of another color," so without a clonal name in addition, *discolor* alone could apply to various color differences.

Var. *flava* Hort.—so many plants have borne the name *flava* that it is a useless name except to indicate a yellow-colored flower. We do not know whether all have been yellows of *Sophronitis coccinea*, or whether other species have been involved. Finding a yellow variant has been so seldom and so valuable to date that growers do not bother to say anything about the vegetative parts of the plants; automatically it is assumed they are forms of *S. coccinea.* There is some indication, in fact, that yellow forms of *S. mantiqueirae* may be more common than yellow *S. coccinea* clones. We do not even know which of the yellow clones today are divisions, and which are different. Of course, *all* should have received a distinctive clonal name when discovered, besides the designation var. *flava.* So far as I can determine, no one in Brazil is concerned with these problems so there is a repeat of the *Cattleya intermedia* var. *aquinii* or *Laelia purpurata* var. *coerulea* confusions described in Volumes I and II. If we are dealing with a yellow form of *S. coccinea,* there has never been mention of the leaves or whether the plant has a normal vegetative appearance for that species (i.e., a red pigmentation pattern). See varieties *lobbii* and *lowii* below.

Var. *gigantea* Hort.—is a term reserved for the large-flowered types discussed above that are presumably tetraploid plants. According to Fowlie, they have been found especially in the Serra do Mar habitat as scattered individual plants. The leaves are as much as 15 cm tall on pseudobulbs of equal length. Flowers can measure at least 7–8 cm in natural spread. Needless to say, these plants deserve individual clonal names when shown. It must also be kept in mind that these may be cultivars of the recently described *Sophronitis bicolor,* noted for the large size of its leaves and the broad, yellow and red-veined lip of the flower.

Var. *lobbii* Hort. *ex* Decker—a plant that was described initially as a clone and therefore the name should have been written 'Lobbii'. It was a very unusual plant with lemon yellow flowers and no longer exists except in the records. Similarly colored clones found today deserve their own cultivar names unless this epithet is changed formally to mean a group of clones together, all having lemon yellow flowers. I am not convinced it was a form of *Sophronitis coccinea* rather than one of the other species. Two features of these individual *lobbii* plants have not yet been commented upon in the literature. Are the yellow-flowered clones found growing among normal red ones? And do they also show the red line along the midrib and around the leaf margin, or do the leaves, like the flowers, lack red pigmentation?

Var. *lowii*—a yellow form that was given an AM when shown at the Royal Horticultural Society on 16 January 1923. It belonged to Pantia Ralli Esq. of Ashtead, Surrey. Although described as a species, then relegated to varietal status, the name should have been a clonal designation 'Lowii'. The original plant no longer survives, and the name should no longer be used (Plate 50).

Var. *pallens*—is called by the Brazilians *abobra,* meaning the color of a local variety of squash, a pale light brick color. Not common, it has not been illustrated and may actually represent plants of one of the other species.

Var. *rossiteriana*—was originally described by Rodrigues in 1877 as a species, but was really only a clonal designation, as indicated by Pabst in 1972. The epithet has also been used horticulturally as a varietal name, but possibly should not even be listed under *Sophronitis coccinea*. The flowers were described as sulfur yellow with obscure orange stripes. The plant was collected by P. Binot near Barbacena in Minas Gerais. That would place its habitat in the Serra da Mantiqueira and, with its rounded, oval pseudobulbs, oval leaves, and tapered, pointed petals, would more likely make it a variety of what today we call *S. mantiqueirae*, not *S. coccinea*. Neither its picture nor its description are adequate to make the decision. Since the original clone no longer exists, the name becomes obsolete, though of historical interest. If it were a clone of *S. mantiqueirae*, the *rossiteriana* epithet would really be the basionym for naming that species. Other plants today with yellow coloration should be given cultivar names under the appropriate species: *S. mantiqueirae* or *S. coccinea*. The group photograph in photo 9 of *Native Orchids of Brazil* shows a good range of colors from red through orange to yellow of *S. coccinea*, but what we should be calling these different colors and forms is still open to critical attention. Certainly each should have its own clonal name, not a varietal designation. Some Brazilians call the yellow color *cor de gemma*, egg yolk color. A plant called *S. grandiflora* var. *rossiteriana* was awarded an AM on 10 January 1899 by the Royal Horticultural Society. The award was raised to an FCC on 26 February 1901. The plant belonged to Sir Trevor Lawrence (Plate 51).

Var. *striata* Hort.—this can also be a confusing category, as there is a tendency for all sophronitis flowers to have darker pigmentation along the vein areas, regardless of the base color. However, this category includes the clones that have a pronounced color difference to the point of being striped in a noteworthy fashion. I do not know of any color illustrations at this time, nor is any mention made of the base color of the flower segments.

Var. *xanthochila* Hort.—plants with flowers having a yellow lip. A confusing category also applied as a varietal name to other species in this genus. Such plants should have distinctive clonal names, not the varietal designation. The epithet should be spelled *xanthocheila*.

The following subspecies epithets have also been used, but they are not validly published, and their relation to the ecotypes listed above is unknown at present.

Subsp. *parviflora*—is illustrated in photo 7 in *Native Orchids of Brazil*. Fowlie has transferred the name to make it a variety of *Sophronitis mantiqueirae*. The Brazilians used the *parviflora* designation to distinguish plants of *S. mantiqueirae* (before it was named as a species in 1968) from the larger-flowered *S. coccinea*.

Subsp. *orgaoensis*—is illustrated in the *Orchid Digest* (1975). This name was used to distinguish a race of *Sophronitis coccinea* from southeast of Nova Friburgo in the Organ Mountains. The plants grew at 1325 m and were stronger, with larger flowers, than those from the Serra do Mar farther south.

Subsp. *espirito-santense*—is for plants from Espírito Santo mountains. They flower from September through November and share certain habitats with *Sophronitis wittigiana,* which flowers in June and July in Brazil.

In 1933 Porto and Brade proposed the generic name *Eunannos* for species with single flowers in the genus *Sophronitis,* reserving *Sophronitis* itself for multiflowered species, namely *S. cernua*. The idea has not been accepted and seems completely unnecessary at present.

Mantiqueira Sophronitis

Brazil

(PLATES 52, 53, 54, 55)

Sophronitis mantiqueirae Fowlie. 1972. *Orchid Digest* 36 (5): 181–194.

Synonyms

Sophronitis coccinea subsp. *mantiqueirae* (Fowlie) Fowlie. 1968. *Orchid Digest* 32(9):272–273.

Sophronitis coccinea var. *parviflora* Hort. 1972. Fowlie in *Orchid Digest* 36(5):181–194.

Before Fowlie's research brought him to a formal naming of the variety as a subspecies of *Sophronitis coccinea* in 1968, Brazilians had distinguished a smaller-flowered form of the scarlet sophronitis for a long time by calling it var. *parviflora*. The *S. coccinea* epithet had for years really been covering the populations of two different species. Realizing that the small-flowered plants were completely different from those of true *S. coccinea,* Fowlie amended the name to full species status as *S. mantiqueirae* (1972), discarding the relationship to *S. coccinea*. He does retain var. *parviflora* as a name for smaller types of this new species (presumably not the result of poor cultivation) and then calls more usual forms *S. mantiqueirae* var. *mantiqueirae,* as is correct by the present rules of nomenclature. The *mantiqueirae* epithet started out as a varietal designation under *S. coccinea,* but without valid publication. There have been a lot of name changes and thrashing about with these sophronitis populations to determine what each name actually stands for and whether or not the names are valid, but as a result we have a better understanding of the whole genus than was ever possible before. The name changes have not been in vain, though all the complications have not yet been completely resolved.

A case in point is the status of *Sophronitis rossiteriana* for which there is an illustration (Plate 51) and type description. Though it has traditionally been considered a variety of *S. coccinea* (please see), the description and flower could well fit plants of *S. mantiqueirae*. If it were this species instead of *S. coccinea,* then the name *rossiteriana,* instead of *mantiqueirae,* would have priority for the name of this species. But who can tell?

The plants of *Sophronitis mantiqueirae* are smaller and have smaller flowers than those of *S. coccinea,* its nearest relative (Plate 52). The leaves lack the red midrib and marginal red pigmentation, and they are oval in shape and curved, not elongated and pointed. The plants flower in the

Brazilian summer months and have a separate habitat area in the Serra da Mantiqueira. The leaves also show red pigmentation on the reverse sides, often in spots toward their bases. The pseudobulbs are fatter and not as long as those of *S. coccinea,* even though they would not be described as spherical. They may also have a red flush, as well as red leaf margins, if the plants are grown in high light. The petals tend to be pointed, longer, and narrower than *S. coccinea* petals, giving the whole flower a wider look than *S. coccinea* flowers, and the flowers are not as high from top to bottom. The lip is elongated to a blunt point and has some distinct yellow veining at its base. Its tip generally shows the same red pigmentation as the whole flower.

The variety *xanthocheila* has an all yellow lip, but this term is also used for forms of *Sophronitis bicolor, S. coccinea,* and *S. acuensis* with all-yellow lips. The plants would, however, ordinarily be flowering in different seasons. Clonal designations should accompany invalid varietal names such as this one.

The *parviflora* types have flowers with a natural spread of 2.5 cm, while more usual forms measure double that. Petals measure 12–14 mm wide × 20–25 mm long, are oval-elliptic in shape. Flattened lips are 18–20 mm wide and 26–32 mm long. Columns are 8–10 mm long, and pedicellate ovaries are 25–35 mm. long.

Since *Sophronitis mantiqueirae* plants are summer flowering in Brazil, they are easy to distinguish there during January and February, as none of the other species are ordinarily blooming then. Fowlie describes a typical habitat as occurring at "4000 to about 6200 ft. [1200 to 1890 m] elevation, growing on small saplings which are usually moss-shrouded (in the creek gulches) or lichen encrusted (on the crest forests)." Although January–February is the main flowering season in Brazil, a secondary blooming may occur in September and October. The seed pods ripen in about 10–11 months in the wild.

A variety *veronica* Hort. from the Pico Crystal region in the western extent of the Mantiqueira mountains has a deficiency of red pigmentation, so the colors range in the salmons and yellows. Fowlie reports that the very small, round pseudobulbs on these plants are a "peculiar bluish-green," the leaves are roundish, and the plants "prefer the vertical trunks of huge old trees just windward of the very crest." He also tells us that the spread on the largest flower of *Sophronitis mantiqueirae* was 10 cm, comparable then to the spread of the largest flowers of *S. coccinea* that have been seen, but not as fully formed because of their narrow, pointed petals.

Winged Sophronitis

Brazil, Paraguay

(FIGURE 2-3; PLATE 56)

Sophronitis pterocarpa Lindley. 1853. *Paxt. Fl. Gard.* 3:11, fig. 239.

Synonyms

Cattleya pterocarpa Beer. 1854. *Prakt. Stud. Fam. Orch.*, p. 209.
Sophronitis cernua var. *pterocarpa* Hort.

The designation of *Sophronitis pterocarpa* has been somewhat of a mystery for a long time. It was illustrated in the original description, and that picture (see Fig. 2-3) clearly shows that the ovary is swollen and winged compared to the typical *S. cernua* flower with which it has long been confused. Some clones of *S. pterocarpa* apparently have more developed wings than others.

Another paper (Kraenzlin, 1911), pointed out to me by Eric Christenson, has apparently been missed until now. It details, with drawing and description, a unique feature of this species, and indeed of the genus. That feature is a fleshy two-parted callus flap at the base of the lip—best seen in a bisected flower. The flap is positioned so that it lies below the anther cap, obscuring access to the stigmatic surface, and, I presume, the nectary. The groove dividing it into two parts must be the channel for guiding a beak or proboscis inside, past the stigma, and beyond to the opening of the nectary. This arrangement insures that any pollinia attached to the beak will contact the stigma upon entering the flower, and the beak would, in turn, detach new pollinia when withdrawn from the flower. The callus structure seems to be particularly noticeable in this species.

Fig. 2-3. *Sophronitis cernua* var. *pterocarpa*. Type illustration of *S. pterocarpa* from *Paxton's Flower Garden* (1853, p. 475).

Some orchidists have continued to list *Sophoronitis pterocarpa* as a separate species, but usually no one has been convinced it really exists. I used to examine every *S. cernua* plant that came my way at shows and finally a *S. pterocarpa* turned up. I photographed it, winged ovaries and all. This plant, the only living one I have seen among dozens of *S. cernua* clones that have come to my attention, was, as expected, listed as *S. cernua* var. *pterocarpa.* We can now consider it a species in its own right. The plant in Plate 56 is a Brazilian clone, photographed in Brazil by Denis Duveen. In his writings Fowlie "excludes" the *S. pterocarpa* concept from *Sophronitis,* an oversight at this point.

The original plant of *Sophronitis pterocarpa* was found by Gardner in the Organ Mountains and was later noted by Cogniaux in Martius "on rotting trees near Mainarde, in the Province of Minas Gerais." Two specimens in the Orchid Herbarium of Oakes Ames are from Paraguay! One is from the Cerro Seahay in the Department of Altos, the other from the Cerro de Acahay. Cogniaux in Martius also cites Paraguayan records for the *S. pterocarpa* forms.

These are unusual locations for what has been considered a Brazilian species and should be reinvestigated in light of our present knowledge. Since such plants have come from scattered locations over a wide area, including Paraguay, we do not know if they represent colonies of plants or only chance clones. Again, this is a problem requiring direct field work and further investigation.

Pygmy Sophronitis

Brazil

(PLATE 57)

Sophronitis pygmaea (Pabst) Withner. *comb. nov.*
Basionym: *Sophronitis coccinea* subsp. *pygmaea* Pabst. 1976. *Bradea* 2(12):70.

Sophronitis pygmaea is the smallest species in the genus. Its globular pseudobulbs and lanceolate leaves scarcely reach 5 cm, usually not more than 3–4 cm. The tiny, deep red flowers are even smaller than some *S. cernua* blooms and have rather pointed segments. Pabst says the flower structure is the same as that of *S. coccinea,* only smaller. In the type dimensions, pseudobulbs measure 5 mm, leaves are 3–4 × 25 mm, sepals are 4 × 10 mm, petals are 8 × 12 mm, the flattened lip is 9 mm across the lateral lobes and 9 mm long. There is only one illustration to date—in the Nov.–Dec. 1977 *Orchid Digest,* p. 235—plus the outline drawing with the type description in *Bradea.*

The only thing we know about *Sophronitis pygmaea* besides the type description is the caption under Duveen's photo in the above-mentioned *Orchid Digest.* He tells us that the species was discovered in the coastal mountains of Espírito Santo by Roberto Kautsky. Since there is no reason to make it a subspecies of *S. coccinea,* I am raising it to species status. It is already known in the trade as *S. pygmaea,* and we still know almost nothing about it from cultivation.

Wittig's Rosy Sophronitis

Brazil

(PLATES 58, 59)

Sophronitis wittigiana Barb. Rodrigues. 1878. *Revist. Hort.*, p. 24.

Synonyms

Sophronitis purpurea Reichenbach f. 1878. *Gard. Chron.*, n.s. 9:462.
Sophronitis rosea Hort. *ex* Gastling. 1884. *Gard. Chron.*, n.s. 21:243.
Sophronitis grandiflora var. *rosea* Hort. 1884. *The Garden* 25:178, 474 and t. 643.
Sophronitis violacea O'Brien. 1886. *Gard. Chron.*, n.s., 18:492.

Some are still confused about this concept of *Sophronitis wittigiana*. Many have insisted that it is a color form of *S. coccinea;* others have said that *S. rosea* and *S. wittigiana* are different forms of pink-flowered sophronitis belonging to two different species. Having investigated this matter, I would like to be counted among those who say that *S. rosea* is a later synonym of *S. wittigiana,* therefore the same species, and that neither epithet has a relationship under the epithet *S. coccinea*. Since the broadly oval foliage and general habit of growth forming a double row of pseudobulbs are distinctive, it cannot belong to the *S. coccinea* concept that has completely different vegetative habits and elongated leaves carrying a red stripe down their midrib.

Besides the unique color of the flower, there is also the noteworthy elongated pedicellate ovary that is 6–8 cm long. It is the only species of sophronitis with this stalk longer than the leaves, and it shows well in the type illustration for *Sophronitis rosea* reproduced here from *The Garden* (1884), a plant flowering in Peacock's collection at Sudbury House, Hammersmith, on 8 February 1884. The species is named after Emile Wittig who first sent plants from the interior of Espírito Santo to London in 1880.

There is some variation in lip color of this species, certain cultivars having more yellow at the base than others. The flowers appear veiny, as often happens with various kinds of sophronitis, with stronger pigmentation over the veins—especially with good light for observation.

The habitat of this species is in the state of Espírito Santo. It occurs, according to Fowlie's data, from the slopes of the Serra Caparão to east of Pedra Azul and about as far north as Mantena at about 1067–1371 m elevations in what he calls "gully swamps." During seasonal drier spells the swamps keep up the humidity around the plants. The plants flower there from April through June; in Hamilton the data show December and January blooming is strongest for cultivation.

Fowlie and others believe that Reichenbach's *Sophronitis purpurea,* also called *S. grandiflora* var. *purpurea,* belongs with this species concept, as was discussed under *S. coccinea*. I find nothing inconsistent with that idea.

A nearly pure white form of this species was found by Hector Gloeden near Pedra Azul, an interesting contrast to the yellow forms of the other species when lacking their dominant red coloration. This would indicate that *Sophronitis wittigiana* has more of a lavender labiate cattleya type of pigmentation system. The lack of anthocyanin pigments then results in white flowers. In contrast, when *S. coccinea* and the other species with their particular combinations of carotenoids and anthocyanins lose the reds, the flowers are yellow from the retained carotenoids. The apparent lack of dominant carotenoids in *S. wittigiana* helps explain its rosy instead of reddish coloration.

The Royal Horticultural Society in the past has considered *Sophronitis rosea* to be a color form of *S. coccinea* and at the same time has registered hybrids under *S. wittigiana* as a separate species.

CHAPTER 3

Miscellaneous Genera Centered in South America, Especially Brazil

Refer to the introductory information in the preface of this volume on the classification of the alliances or series of the Laeliinae to which the various genera may be assigned. Genera centered in the Caribbean, México, and Central America will be included in Volumes IV, V, or VI.

DESCRIPTION OF GENERA AND SPECIES

Constantia
Isabelia
Lanium
Leptotes
Loefgrenianthus
Neolauchea
Pinelia
Pseudolaelia
Pygmaeorchis
Reichenbachanthus
Renata
Sophronitella

CONSTANTIA

This two-leaved genus was proposed by J. Barbosa Rodrigues in 1877. It is named after his wife, Constança, and at present there are five species, two of which are newly published. Cogniaux combined *Constantia* with *Sophronitis* in 1898, making it a two-leaved subgenus under *Sophronitis*. Subsequent orchidologists have ignored this idea and consider *Constantia* to be a good genus in its own right. Both genera have eight pollinia. The only species we ordinarily see in cultivation is *Constantia cipoensis*, though recently *C. rupestris* has also been available.

The genus is characterized by having eight pollinia, the same as *Sophronitis*, and a small lip that is undivided and not much larger than the sepals, which are about twice the width of the petals. Previously, it has been stated that the lip of the flower is directly attached to the base of the column, no sac-like or spur-like structure being formed at the juncture as it is in other genera. This now seems in question since the drawings of Miranda, illustrating his two new species, show a distinct fusion of the base of the lip with a somewhat footed column. There is a sac-like, infracolumnar enlargement that is most likely a nectary. The flowers of the other species should be reexamined on this point. (See Fig. 3-3.)

In 1983 Ann Jesup registered a hybrid, the first for this genus, between *Sophronitis cernua* and *Constantia cipoensis*. She called it *Conphronitis* Cerci.

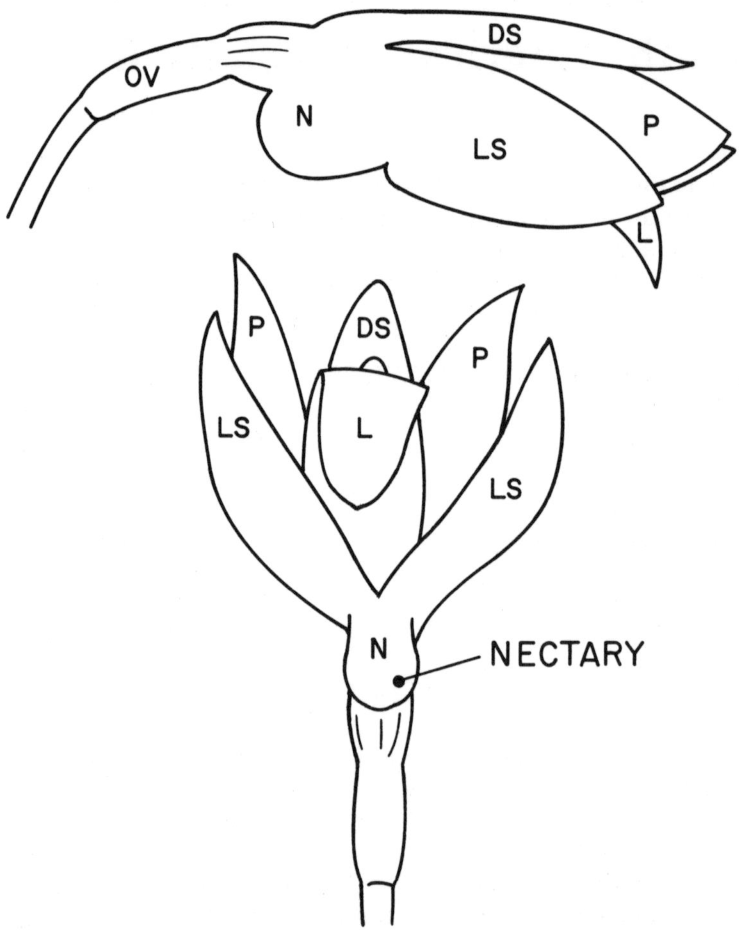

Fig. 3-1. Diagram showing lip fusion with the column and lateral sepals to form the nectary in some species of the Laeliinae.

Southern Constantia

Brazil

(FIGURE 3-2)

Constantia australis (Cogniaux) Porto and Brade. 1935. *Arq. Inst. Biol. Veg.* 2(2):208.

Synonym

Sophronitis australis Cogniaux. 1898–1902. Martius' *Flora Brasiliensis* 3(5):322.

Constantia australis was described first as a *Sophronitis* and then was transferred to *Constantia* by Porto and Brade. The plant was collected in Santa Catarina growing on rocks near Desterro and flowered in March. It apparently is limited to that area in its distribution and has since been collected only two or three times. A comparison of the lip and other flower parts with the other species shows it to be distinct in its proportions. It has larger flowers with a smaller lip than those of *C. rupestris* and is reported to have a pale pink to white coloration. The lip has a different configuration than that of *C. cipoensis*. The following analysis is taken from Cogniaux's original description.

The reddish, off-center pseudobulbs are globose 6–9 mm in diameter, with two rigid, small, gray-green leaves 5–7 mm wide and 4–6 mm long, somewhat kidney-shaped with an indented apex. The leaf veins are ribbed and parallel. The flower stalk between the leaves is only 3–4 mm long with a single flower. Sepals measure 3–3.5 × 7–8 mm, petals are ½–⅓ the width of the sepals, and the narrowly obovate lip with a blunt tip measures 3 mm wide and 5 mm long. There are two little lateral points on the basal margins. The column is 5 mm and is free from the lip.

Though poorly known, *Constantia australis* would seem to be a distinct species. To my knowledge, it has never been grown nor photographed.

Fig. 3-2. *Constantia australis*. Lip outline comparison with other *Constantia* species. After Pabst and Dungs (1975, 1977).

Cipo Constantia

Brazil

(FIGURE 3-2; PLATE 60)

Constantia cipoensis Porto and Brade. 1935. *Arq. Inst. Bot. Veg.* 2(2):208.

Many have seen *Constantia cipoensis* in cultivation, as many Brazilian nurseries have offered it for sale. The plants are usually shipped as they grow, on a branch cut from the stem of a *Vellozia* from the Serra do Cipo in Minas Gerais at 1400 m. The *Vellozia* immediately indicates the need for a cool, dry environment for growing this species. Several plants have been awarded by the American Orchid Society for culture, reflecting the size of the original plant cut off on its *Vellozia* host. It also has received a CBR and a CHM. The plants flower here in October or November.

The cultivar with largest flowers, 'Roll's Choice' CHM, had a natural spread of 25 mm. The sepals were 3 × 1.5 mm, the petals 4 × 15 mm, and the lip 6 × 15 mm. The flowers were white and, in the photo in the 1987 *Awards Quarterly*, show a longer pedicellate ovary and flower stalk than the other species would have. The dull, grey-green, pointed leaves with dark green midrib are a bit longer than they are broad, the opposite of the other species. They measure 5–7 mm wide × 6–8 mm long. The pseudobulbs are round, flattened apically, and are about 8 mm across. They are usually flattened on their sides from growing compressed so closely together. The lip is fused only at the base of the column and has three yellow callus enlargements on the disc. The column is 4 mm long and 3 mm broad. The species can flower at odd times rather than evenly over a large clump.

Maria Cristina's Constantia

Brazil

(FIGURE 3-3; PLATE 61)

Constantia cristinae Miranda. 1991. *Albertoa* 3(10):93–100.

Published in December 1991, *Constantia cristinae* is a rupicolous species that was found two years earlier near Diamantina in the State of Minas Gerais. It was discovered by Maria Cristina Miranda and named after her by her husband Francisco, horticulturist and orchidologist of Orquidário Boa Vista in Rio de Janeiro. The small, tightly clustered plants are similar to those of the other species but, unlike previously described species, this one has intense rose-colored flowers with a purple column and darker purple anther cap. Miranda states that the plants grow on the vertical sides of large blocks of stone in places that receive some protection from the sun, and they are able to withstand the rigors of a seasonally very dry climate.

The pointed, asymmetrical leaves are carinate and slightly tuberculed in crest-like lines, and the leaf base is somewhat heart-shaped as it is attached around the top of the pseudobulbs. The pseudobulbs are a little flattened diagonally so the two leaves are off center, a characteristic that seems to occur in all the species. The pseudobulbs are small, measuring 1–1.1 cm; the leaves vary from 1.5–4 mm in length. The flower stalk is very short, not as long as the 2 mm floral sheath, and the flowers have a natural spread of 1.3 cm. The pedicellate ovary is about 1 cm long and curves so that the flowers are displayed horizontally.

The sepals are lanceolate, pointed, and a little reflexed and united at their bases around the bulge produced by the nectary. They measure 4.5 × 9 mm and are larger than the petals. The petals are oblanceolate, pointed, and are 3.8 × 9 mm. The lip has elevated, parallel sides at the base with the midlobe wider, rounded, slightly undulate on the edges, and with three, bright yellow, verrucose keels. It measures 5 × 7.5 mm. The column is 5.5 mm long, and the stigma is semicircular in outline. The flowers are fleshy in substance.

The plants had been discovered in 1989, but without flowers, and were finally found in bloom in late August in their natural habitat. Seed pods were not observed on any of the plants and are still unknown. They seem most like *Constantia cipoensis* in their vegetative size and other characteristics, but the flowers are readily distinguished.

Miranda suggests trying to grow these tiny plants on the bottom of an upturned flowerpot. This would be a good way to duplicate their rock habitat in cultivation. They will be a challenge for even the experienced grower. If seeds are ever available we might start sterile flask culture as a way of growing miniature orchids—using a long-term inorganic medium without sugar or other organic supplements that would readily mold. After all, many have observed orchid seedlings flowering in the flask. Oncidiums, broughtonias, phalaenopsis, miniature cattleyas, and still other plants of much larger stature than constantias can flower *in vitro*; why not these smaller species?

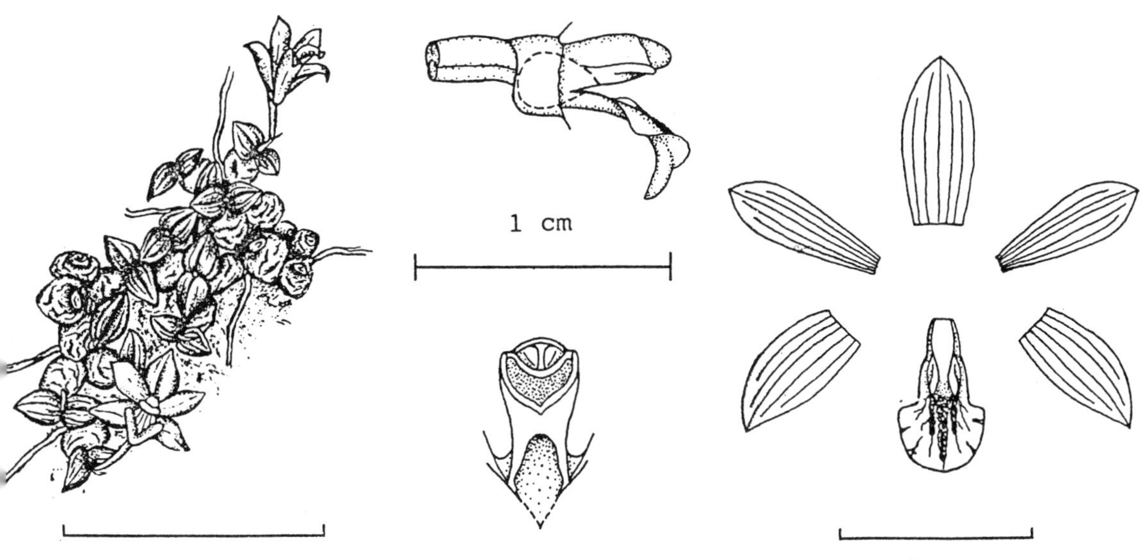

Fig. 3-3. *Constantia cristinae*. Drawing of vegetative habit and floral details. Permission of the author.

Microscopic Constantia

Brazil

(FIGURE 3-4; PLATE 62)

Constantia microscopica Miranda. 1991. *Albertoa* 3(10):93–100.

This new species is called *Constantia microscopica* because of its tiny size, the smallest in the genus. It was found in 1986 by Rolf Grantsau, and Francisco Miranda collected and described the type specimens in 1991. Like the plants of *C. cristinae,* these also grow on large blocks of rock in the vicinity of Diamantina in the State of Minas Gerais. They like vertical surfaces protected from the sun.

In addition to their tiny size, they differ from their congeners by having leathery leaves that are densely tuberculated. The flowers are also coriaceous on a comparatively tall stalk, the tough sepals and petals persisting on the rounded seed pod that is covered in verrucosities. The infracolumnar nectary is comparatively large, and the column is unusually short with a semicircular stigma.

The pseudobulbs are flushed with purple, are rounded but typically skewed off-center, and measure about 4 × 5 mm. There are usually two, rarely three, leaves per pseudobulb. They have a prominent midrib, are slightly convex with distinct verrucosities on their upper surfaces, and have a rugose surface on the lower. They measure 5–6 mm in length. The floral sheath is pointed, about 2 mm high, and the flower stalk has three nodes with bracts and reaches 2.5 cm height. The sepals and petals are pinkish or rose-colored, the column is greenish yellow, the anther cap is purple, and the lip is white.

The ovate, pointed sepals measure 3 × 4.5 mm and the lanceolate petals are smaller, 1 × 3.5 mm. The base of the lip has fleshy, raised crests along the sides, and the midlobe is reduced to a fleshy knob. There are two

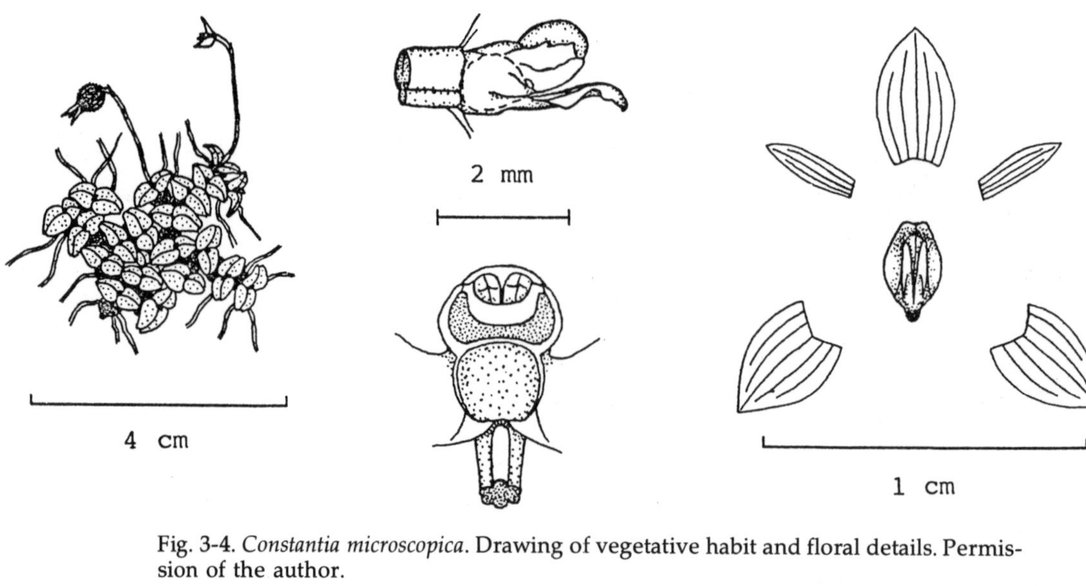

Fig. 3-4. *Constantia microscopica*. Drawing of vegetative habit and floral details. Permission of the author.

fleshy callus keels with a small central vein. The lip measures 2.3 × 3.5 mm; the column is 2.2 mm long.

Constantia microscopica plants are unquestionably the smallest of all the Laeliinae. They can also compete with some of the tiny pleurothallids for being among the smallest orchids of the New World. One can only speculate about the pollinator of such tiny flowers and how the species ever evolved. It will be fascinating to cultivate whenever plants become available.

Rock-dwelling Constantia

Brazil

(FIGURE 3-2; PLATE 63)

Constantia rupestris Barb. Rodrigues. 1877. *Gen. Sp. Orch. Nov.* 1:79.

Synonym

Sophronitis rupestris (Barb. Rodrigues) Cogniaux 1898–1902. Martius' *Flora Brasiliensis* 3(5):322.

Constantia rupestris is the type species for the genus. The pseudobulbs are short and pear-shaped but are usually strongly compressed from growing so close together. They measure 8–10 mm wide and 10–15 mm high. Their color is pale green becoming brownish in color with a large circular leaf scar after the leaves are gone. The leaves are broader than long, 9–10 mm wide × 7–8 mm long, bright green and shiny, and paler on the back. They are somewhat kidney-shaped, with a heart-shaped base and an indented apex.

A single flower is produced on a stalk 10 mm long with a little floral sheath about 1–2 mm long. The flowers face down a little and are a greenish white. The plate in Pabst and Dungs (1:222) is a watercolor painting of this species and shows a purple midrib stripe on the petals, a purple anther cap, and purple edges on the column. The sepals measure 3 × 10–12 mm, the somewhat recurved petals are 1.5–2 × 10–12 mm, and they are narrower below and overall compared to the sepals. The tongue-shaped, rather oblong lip has a strongly recurved, slightly pointed apex, and there is yellow toward the base and on the disc. It is 3 mm broad × 8–10 mm long. The column is 5–6 mm long and is fused at the base to the lip.

This description is taken from the original by Rodrigues in 1877, and little can be added today as this species is almost unknown, even in Brazil. The type specimen came from trees in a rocky, arid location in the Serra de Jacarepagua in Rio de Janeiro state, and it was blooming in March. Plants have also been found near Migui in Espírito Santo. At Kew is a copy of the colored plate #245 of this orchid from Rodrigues's original in his *Iconographie* (1868–1873).

ISABELIA

The genus was founded by Barbosa Rodrigues, a Brazilian botanist. He may have missed out in the priority of naming several of the showy cattleyas and laelias from Brazil (mentioned in the first two volumes of this series), but he certainly managed to name many of these minor genera and their miniature species. On the other hand, Cogniaux missed this genus and several other species when he wrote his definitive account for Martius' *Flora Brasiliensis* that presumably included all of Rodrigues's species. But *Isabelia* is not listed there. It is, of course, listed in the modern *Orchidaceae Brasilienses* of Pabst and Dungs. The name honors Brazilian Princess Donna Isabel, Comtessa d'Eu, and is spelled with only one *l*.

In this genus the base of the lip fuses with the column base to form a tiny sac-like nectary. The flowers have eight pollinia bound in pairs by the caudicles.

The natural hybrid *Isanitella* is described below.

The Princess Isabelia

Brazil

(FIGURE 3-5; PLATES 64, 65)

Isabelia virginalis Barb. Rodrigues. 1877. *Gen. Sp. Orch. Nov.* 1:75–76.

Plants of *Isabelia* are best grown mounted on a tree branch or slab so that the miniature can form a large cluster of rhizomes with small pseudobulbs and remain undisturbed for a long period. The 1 cm pseudobulbs are a few millimeters apart on the creeping stems. I have never ceased to be fascinated by the bracts around the rhizomes and pseudobulbs of this species. They are a most remarkable openwork fabric made of crisscrossed and reticulated veins and fibers that make it appear as though each growth were wrapped in a miniature burlap textile. It would make an interesting developmental study to find out just how these bracts and fibers are formed as the new leads mature. The pseudobulbs are completely enclosed by the bracts, and each possesses a soft but needle-like leaf that may reach 4–6 cm in length. Old pseudobulbs may be flushed with red.

A well-established clump of plants may produce up to 100 or more new growths, each with a single flower about 1.3–1.6 cm across. The sepals and petals look similar, but the petals are about half the width. The narrow base of the unlobed, simple lip is fused to the base of the column. The flowers show up as pearly white with pink flushes and overtones, especially on the sepals. The lip is white also but may have two dark purple areas on the sides deep in the throat. The entrance to the nectary spur may be seen between them. The column is maroon-topped with a mahogany-colored anther cap, and it is definitely set off by the whiteness of the flower around it.

I have found nothing written about the pollinators of this species or any other of these small orchids. It has a wide distribution in the central and southern regions of Brazil. Originally collected near Caldas and flowering in April, this species flowers mostly in November in cultivation. The original plants were collected in the Organ Mountains by Gardner in March 1837 and were thought to be maxillarias. The pollinia showed that idea was incorrect. Though Lindley was uncertain of its placement in classification, he thought it belonged between *Leptotes* and *Sophronitis*. It was introduced to cultivation by K. Grossman, who sent plants to the Berlin Botanic Garden in 1904. The owners of various clones have been awarded cultural certificates by the American Orchid Society for well-grown specimens.

A firsthand account by Professor Milton Leinig (1971) of the plants growing in the crevices and debris pockets of eroded sandstone outcroppings in Vila Velha near Ponta Grossa in Paraná is fascinating. It grows there in association with *Sophronitella violacea,* another rock inhabitant, along with plants of the natural hybrid between the two species that was named *Isanitella pabstii* by Dr. Leinig. It is, so far, the only hybrid involving either genus. Professor Leinig of Curatiba, long a student of Brazilian miniature orchids, found the plant and flowers to be intermediate between its two parents. The flowers are a uniform lavender pink with a bit of white on the disc. The drawings from his article are reproduced here. The lip architecture and the development of the nectary are especially noteworthy. The nectaries are simple versions of the more fully formed cuniculus nectaries in the genus *Brassavola.*

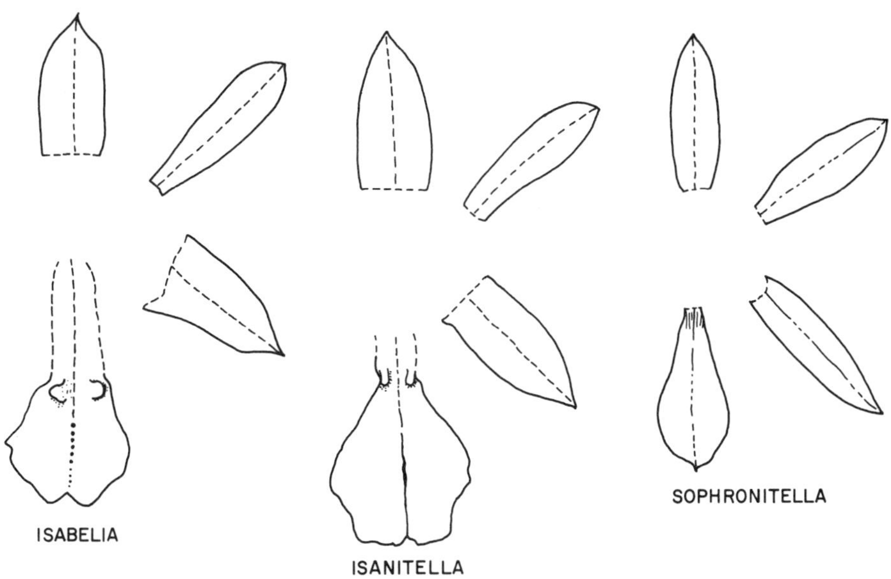

Fig. 3-5. *Isanitella pabstii.* Outline drawings comparing flowers of parents and hybrid. After Leinig (1971). Courtesy of the American Orchid Society.

LANIUM

Lanium, meaning wooly in Latin, was set up by Bentham in 1881 as a separate genus, though Lindley had originated the use of that epithet for a section of the genus *Epidendrum* in 1841. Bentham declared it to be a distinctive taxon worthy of generic status because the backs of the sepals, the stalks or pedicels of the flowers, and the fruits of these species were covered with a fine woolly tomentum or hair; also, the anthers had unequally sized pollinia, and the individual flower stalks showed a unique bend. The bends make the buds appear like "little birds." Otherwise, the flowers resemble those of small, undistinguished epidendrums with greenish or pinkish flowers in terminal racemes.

The lip attaches directly to the base of the column and is undivided in all the species. It has a scoop-like shape with pointed tip. Also, as usual in several of these small orchids, the sepals are two or three times wider than the petals. In all species the sepals are triple-veined and the petals have a single midrib vein. Also, the petals tend to twist so that they have a horizontal orientation with the rest of the flower, reminiscent of *Cattleya walkeriana* flowers.

There are four pollinia, the same count as in *Cattleya, Encyclia,* or *Epidendrum,* but the two central ones are compressed and the two outer are more elongated and comma-shaped. The lip is uppermost in the flowers since the flowers are not resupinate. There is a vegetative variability in the plants collected from different localities so that a certain confusion exists about the validity of some species and the exact parameters of their characteristics. Dressler (1967) has the view that *Lanium* should be a sectional name under *Epidendrum,* going back to Lindley's idea, but most today consider it a separate genus.

Little Bird Lanium

Bolivia, Brazil, Peru

(PLATES 66, 67)

Lanium avicula (Lindley) Bentham. 1881. *Hooker f. Icones Plantarum.* 14:25, t. 1334.

Synonyms

Epidendrum avicula Lindley. 1841. *Hook. Jour. Bot.* 3:85. (Misprinted as *avicule*).
Lanium berkeleyi Rolfe. 1894. *Kew Bull.,* p. 392; Appendix II, p.45.
Lanium avicula var. *longifolium* Cogniaux. 1898. Martius' *Flora Brasiliensis* 3(5):26.
Epidendrum longifolium. 1908. Lehmann specimen #245–08 from Hort. Kew via Brussels Bot. Gard. *Nomen.*
Epidendrum lanioides Schlechter. 1913. *Fedde Repert. Spec. Nov.* 12:489.

These are creeping plants that tend to form large clumps or clusters that may produce hundreds of blooms. One American Orchid Society awarded specimen clone, 'Mt. Vernon', produced 2250 flowers on 90 inflorescences held above the leaves. In Latin *avicula* means "small birds," which refers to the appearance of the flowers. The sepals and petals are brown, the lips green and spade-shaped. The dorsal sepal reflexes in the flowers, an undesirable trait when it comes to judging most orchid flowers, but far from uncommon among the cattleyas or laelias. In fact, it is nearly a stable genetic characteristic for several species besides this miniature. The individual flowers are 1.4 cm across, sepals measuring about 6–7 mm long, and the lip similarly sized. The column is yellow-green and may have some fine red speckles.

The pseudobulbs are ovoid, 2–3 cm tall, and the two leaves can measure up to 15 mm wide and 25 mm long. The growths are separated on the rhizomes so that the mass readily tangles together as the plant grows around a branch. Brieger, Maatsch, and Senghas in *Schlechter III* distinguish several Brazilian varieties depending on the shape of the leaves—band-like, oval, rounded, thicker, thinner, and so on—so the plants are somewhat variable depending upon their specific habitat in the various Brazilian states.

In *Schlechter III*, Brieger makes *Lanium berkeleyi*, originating in the state of Ceara, one of these leaf variants. In Pabst and Dungs (1975, 1977), the outline drawing of *L. avicula* is very similar to *L. berkeleyi* in shape and size, but the lip has three central veins and only a small raised area at base. Hair does not show on the sepals in their drawing.

The plants flower from September through November according to the Hamilton data. The original plants came from Rio de Janeiro.

Berkeley's Lanium

Brazil

(FIGURE 3-6)

Lanium berkeleyi Rolfe. 1894. *Kew Bull.*, p. 392; Appendix II, p. 45, 1895.

This *Lanium* was found growing among the roots of a plant of *Cattleya guttata* sent to Major General E. S. Berkeley of Southampton. It first flowered in 1891 and was distinguished from *Lanium avicula* by its longer, narrower pseudobulbs and leaves, and less-branched raceme. The pseudobulbs are 2–3 cm tall, and the foliage consists of two or three erect stiff leaves that are 6–11 mm wide × 20–37 mm long. The laxly flowered inflorescence is graceful with 6–9 flowers and is longer than the leaves and

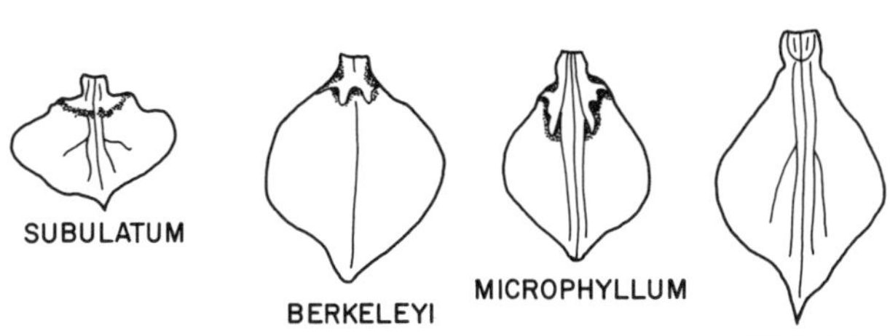

Fig. 3-6. *Lanium berkeleyi*. Outline drawing of lip with other *Lanium* species for comparison. After Pabst and Dungs (1975, 1977) and other authors.

hairy. The small flowers are green with minute red-brown dots. The lip is shorter than the lateral sepals and has three central thickened veins with a bidentate callus at the base.

The drawing in Pabst and Dungs shows that the sepals are hairy-edged and that the lip has definite callus area at the base with one central vein. Otherwise, the flowers are similar to those of *L. avicula*. Flowering, according to Hamilton, is also similar—a fall bloomer. With more information about this species, these plants may turn out to be variants of *L. avicula* as some have stated.

Creeping Lanium

Cuba, Bolivia, Brazil, Colombia, Ecuador, Guyana, Peru, Suriname, Venezuela

(FIGURE 3-7)

Lanium microphyllum (Lindley) Bentham. 1881. In *Hooker f. Icones Plantarum* 14:24, t. 1334.

Synonyms

Epidendrum microphyllum Lindley. 1840. *Hook. Jour. Bot.* 3:85.
Epidendrum sarcophyllum Focke. 1849. *Tijdshr. Nat. Wetensch.* 4:67.
Lanium colombianum Schlechter. 1921. *Fedde Repert. Spec. Nov.* 7:155. Atlas, t. 211.
Lanium ecuadorense Schlechter. 1921. *Fedde Repert. Spec. Nov.* 8:89. Atlas, t. 368.
Lanium peruvianum Schlechter. 1921. *Fedde. Repert. Spec. Nov.* 9:97. Atlas, t. 481.
Lanium hioramii (Acuna and Roig) H. Dietrich. 1980. *Wiss. Zeitschr. Friedrich-Schiller Univ. (Jena), Math,-Naturwiss. Reihe* 29, 4:522.

Plants of *Lanium microphyllum* appear as a creeping, twisted mass of miniature rhizomes and growths with short, fusiform secondary stems only 2–6 mm long; it is not pseudobulbous, bearing multiple foliaceous bracts and small, stiff leaves only 15–25 mm long. The racemes have 5–8 fragrant flowers 1 cm in size. The sepals and petals are reflexed back from the column and are a greenish yellow with flushes of pink or purple. In Pabst and Dungs (1975, 1977), drawing #804 shows the lip with three central veins and a callus area of thickened veins toward the base. Hamilton data show peak flowering in March.

This species has a wide distribution. It is also an unusual distribution, since it is found in Cuba as well as most South American countries but not in Central America. There is also the possibility that the Cuban plants are a separate species. According to Dodson and Dodson (1980) in Ecuador it is an "epiphyte on the upper limbs of large trees in montane wet forest." Despite its widespread occurrence, it is seldom seen in shows or collections, even as a curiosity.

Fig. 3-7. *Lanium microphyllum.* Plate 538 from *Icones Plantarum Tropicarum* (Vásquez and Dodson, 1982). Courtesy of the artist and of the Marie Selby Botanical Garden.

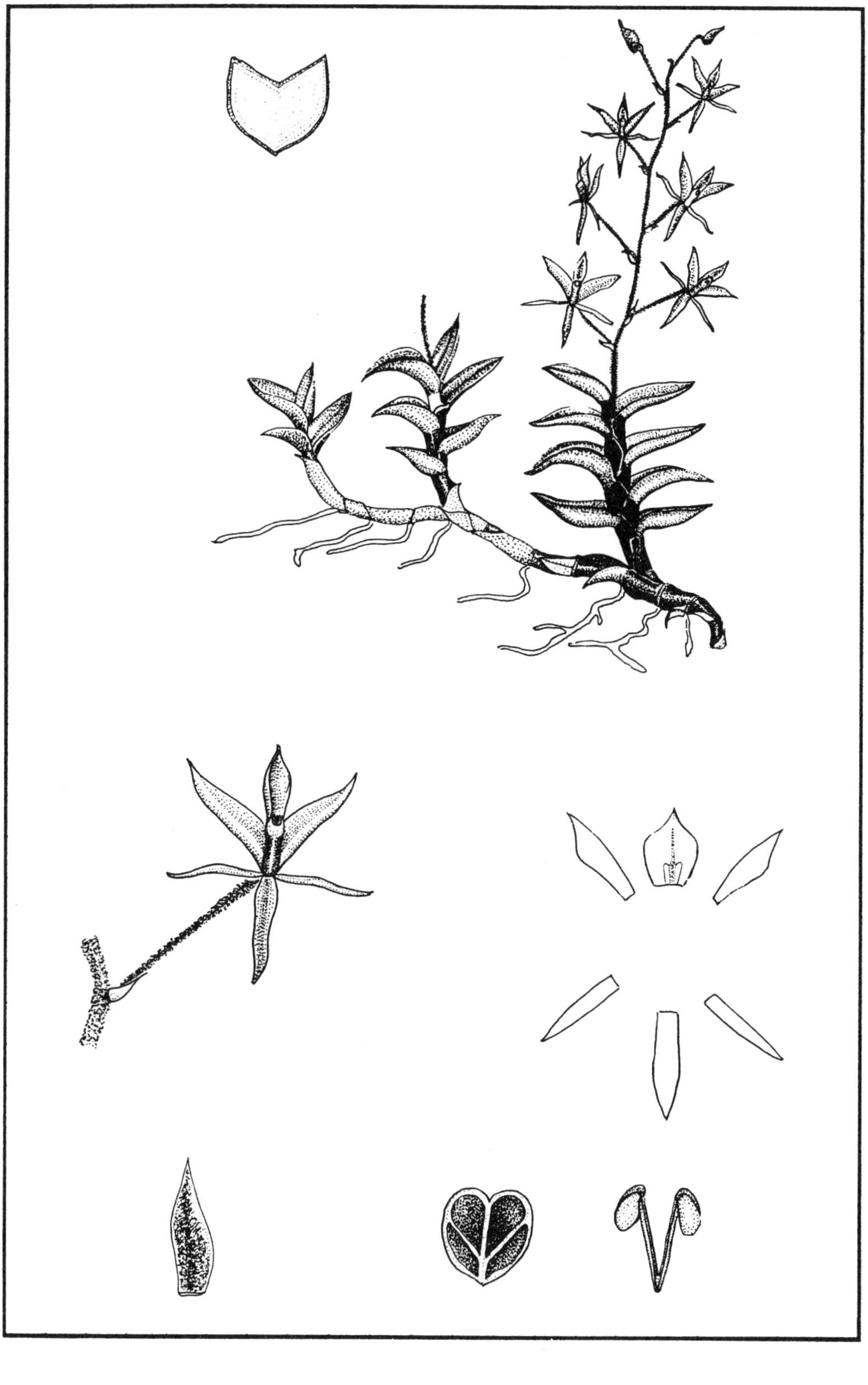

Awl-leaved Lanium

Lanium subulatum Rolfe. 1896. *Kew Bull.,* p. 46; Appendix II, 54, 1897.

Synonym

Epidendrum stiliferum Dressler. 1967. *Brittonia* 19:242. *non E. subulatum* Koenig, 1791, *nec* Swartz, 1788.

This species, imported with a lot of *Cattleya dolosa,* has fat, ovoid pseudobulbs that are two-leaved. The pseudobulbs are 12–18 mm long, and the leathery, linear leaves, stiff, pointed and upright, grow to 45 mm. The wide-branched, open panicles of many tiny green flowers are held well above the leaves. The petals and sepals are only 3 mm long, and the lip is only 2 mm in length. In Brazil the plants come from Minas Gerais and Goias.

The drawing #805 in Pabst and Dungs (1975, 1977) shows that the lip is transversely wider and shorter than in other species, and there are three central veins with two more projecting outwardly (Fig. 3-6). The callus area is semicircular at the lip base. In the *Schlechter III* photo, page 565, this lip shape is not immediately apparent. Dressler (1967) believes that the Peruvian collections may represent a new species. They were collected from the canyon of the Rio Marañón.

LEPTOTES

Although the generic name comes from the Greek word meaning thin and tender or delicate, the leaves of this taxon tend to be somewhat succulent and tough, though they are comparatively small, channeled, and terete. It was named, nevertheless, by Lindley in 1833. The pollinia are of an unusual type, not like those of any other genus in the subtribe. There are four smaller pollinia in two pairs and then one large pair in the center, all held together with very short caudicles (Fig. 3-8). Each large pollinium of

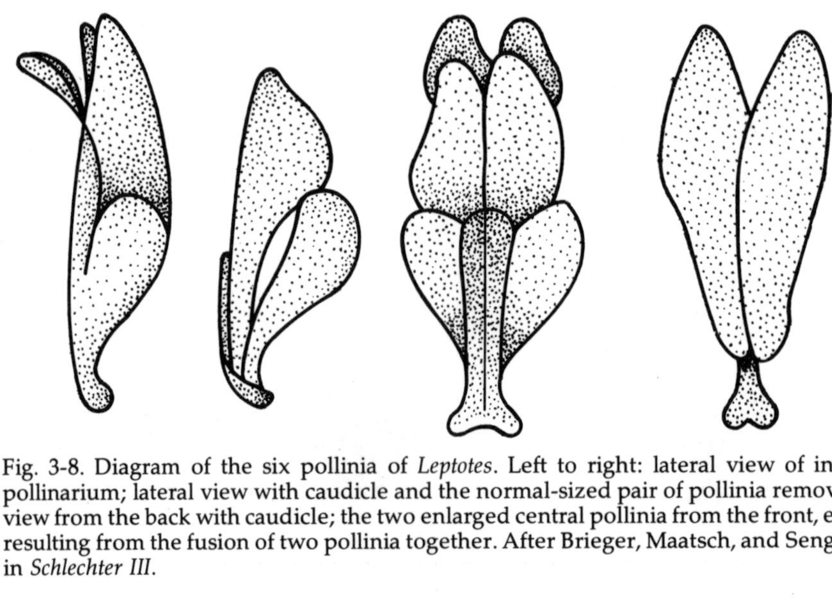

Fig. 3-8. Diagram of the six pollinia of *Leptotes*. Left to right: lateral view of intact pollinarium; lateral view with caudicle and the normal-sized pair of pollinia removed; view from the back with caudicle; the two enlarged central pollinia from the front, each resulting from the fusion of two pollinia together. After Brieger, Maatsch, and Senghas in *Schlechter III*.

the central pair represents a fusion of two others to account for the basic number of eight, though by actual count the number is six. The fruit of *Leptotes* is succulent and fragrant like that of *Vanilla,* an observation described originally by Barbosa Rodrigues.

The plants seem much affected by the amount of light and moisture available during their growth, some plants being short and stocky while others of the same species may be more slender and elongated. In my experience the plants prefer to be mounted with their roots exposed, though it may take a season or two to get them established.

Bicolored Leptotes

Brazil, Paraguay

(PLATES 68, 69)

Leptotes bicolor Lindley. 1833. *Bot. Reg.* 19, t. 1625.

Synonym

Tetramicra bicolor (Lindley) Bentham. 1881. *Jour. Linn. Soc. London, Bot.* 18:314.

Veitch says that *Leptotes bicolor* was first introduced from the Organ Mountains by Mrs. Arnold Harrison of Liverpool in about 1831 or 1832, and was later sent by Gardner from the same region to the Woburn collection. It flowered first in England in February 1839. The plants require intermediate conditions, growing well with Brazilian cattleyas, and like to be mounted on cork with a little moss or be in a basket with osmunda. The seed pods are used for flavoring by local people.

The creeping rhizomes produce clustered short stems closely sheathed in white, dry bracts. The terete leaves are distinctly channeled on the upper surface and are 5–10 cm long. The flower stalk is 1–3 cm long, and the natural spread of the 1–3 flowers that are usually produced is about 4–5 cm. The flowers are a little unkempt with sepals and petals curving forward, or in any other direction, so that the flower width and shape is variable.

The plants are floriferous, and the American Orchid Society has given several cultural awards, some to plants with more than 100 blossoms. The flowers vary from white to a pinkish lavender or purple, some having only a blush of color. There is a rosy magenta on most of the lip, and the column is also colored. The edges of the pointed midlobe usually curl under to produce a tubular effect, and the small lateral lobes project forward on either side of the column.

Several color variations have been formally named, indicating the color variation of the flowers and also the size variation of different clones from different habitats. Among these are var. *alba,* var. *brevis,* var. *concolor,* and var. *glaucophylla.* Plants of the species are found in the states of Minas Gerais, Rio de Janeiro, Espírito Santo, São Paulo, and Santa Catarina—in other words, they have a wide distribution in Brazil. The species is also found in Paraguay. One peloric form was described in the older literature, but it is not known at present.

São Paulo Leptotes

Brazil

(PLATES 70, 71)

Leptotes pauloensis Hoehne. 1933. *Bol. Agricola de São Paulo* 34:620–621.

Little is known of *Leptotes pauloensis* but its type description, and it has been confused as a synonym of *L. tenuis,* which it is not—and vice versa. In 1989–1990 it suddenly became available from one of the Brazilian nurseries, and now several seem to have it. The native collectors, the *caboclos,* must have discovered a new location with different orchids to garner.

The plants form clumps of short, upright, terete, channeled dark green leaves with red-flushed bases about 3–4 cm high. The short purple flower stalks scarcely project, each with a single flower. The lip is three-lobed, and the midlobe is somewhat fan-shaped or semicircular with a rounded rough and wavy margin. The flowers are pale pink to white and have rose to violet veining on the lip. The plants seem to grow easily when mounted on a piece of tree fern, requiring intermediate conditions. They flower during the late summer and fall in cultivation. There is one record of March blooming in Hamilton's data, and Orchidanica Company, Oakland, California, flowered it in July 1990 for the Santa Barbara Orchid Fair.

Serrate Leptotes

Brazil

(PLATE 72)

Leptotes serrulata Lindley. 1838. *Sert. Orch.,* t. 11.

Synonym

Epidendrum ficoide. 1838. Name on Descourtilz drawing, plate 28, p. 63. Published by Lindley as above.

Plants of *Leptotes serrulata* are unique in the genus for their often bifoliate thick-leaved growths, their larger, usually white flowers, and the well-rounded and serrated edges of the lateral lobes of the lip. The plate in *Sertum Orchidaceum* shows five flowers produced per growth instead of the usual three or fewer of *L. bicolor,* and they are on stems long enough to exceed the leaves. In *Schlechter III* (9:563) is a detailed photo of this species mistakenly labeled "*unicolor*" that shows a good comparison with the *L. bicolor* illustrated above it. One growth shows seven flowers. Several past authors have made this species synonymous with *L. bicolor,* but it will be kept here as a separate species for the above differences, hoping that new observations may resolve its species status.

The foliage is a glaucous green color and has red-purple speckled markings, especially on the underside. Lindley quotes from the description accompanying Descourtilz's plate:

> This charming plant is especially remarkable for the sweet odour of the Lilac which its flowers exhale. It is found in blossom in the month of December, on the trunk of Cedrela trees, in the ancient Rocas of Brazil, where, without any sign of suffering, it survives the conflagrations that destroy so many other plants. I also found it in plenty in the district of Upper Macahe, and in that of Ilha Grande, where however it is more rare.

Lindley goes on to describe the flowers as having enamel-white sepals and petals, the sepals "ribbon-shaped, rather broad," and the petals "narrower and thinner." The lip at the base has a white center from which radiate lines of "brilliant lilac, and is afterwards dilated into an ovate pointed or lanceolate limb of a beautiful white." I would add that the midlobes of *Leptotes bicolor* usually lack the veining and are flushed with the color.

One plant under this name with the cultivar name of 'Ruth' was given a CBM in 1965 by the American Orchid Society. Otherwise, not much seems to be known about the species. At the same time, we should keep an open mind and be on the lookout for it as distinct from *Leptotes bicolor*.

Slender-leaved Leptotes

Brazil

(PLATE 73)

Leptotes tenuis Reichenbach f. 1865. *Hamb. Gartenz. Blumenzeit.* 21:296.

Synonym

Leptotes minuta Rolfe. 1889. *Gard. Chron.*, 3rd s., 6:323.

The clumped, terete foliage of this species is tightly clustered, comparatively slim and long, but otherwise recognizable as a *Leptotes*. It appears similar to the foliage of *L. unicolor*, but the flowers are completely different so that the species should no longer be confused with one another. The thin, wire-like flower stalks project out from the base of the leaf clump and hang down, each with a single, small flower.

The flowers are creamy colored with a spot of purple on the disc of the lip. The stigma is outlined in violet. The species was discovered by Blunt, a collector for Low and Company in England. One clone, called 'Lemon Plum', was granted a CBR by the American Orchid Society. It belongs to Larry Moscovitz of Orchidanica Company, Oakland, California. The plants flower in summer, July or August.

Concolored Leptotes

Brazil

(PLATES 74, 75)

Leptotes unicolor Barb. Rodrigues. 1877. *Gen. Sp. Orch. Nov.* 1:74.

Synonym

Leptotes paranaensis Barb. Rodrigues. 1882. *Gen. Sp. Orch. Nov.* 2:163.

The small pendulous plants of *Leptotes unicolor* tend to produce two fragrant flowers on stalks shorter than the leaves, and they are a pale rose-lilac, sometimes almost white, overall. Leaves vary from 2–5 cm long, are 3–8 mm wide, and are sometimes very succulent for their size. As with other species of *Leptotes,* the petals are narrower than the sepals and the flowers face down, never opening widely. The flower parts all measure about 2–2.5 cm long, and the three-lobed lip is only 8–10 mm wide. It is narrowly oval and sharply pointed. The plants may be found in Minas Gerais and flower in Brazil in May and June. That translates into November and December, according to Hamilton's data, when cultivated in North America.

Leptotes paranaensis, made a synonym of *L. unicolor,* by Pabst and Dungs was described by Rodrigues as having only single flowers, more elongated and slender foliage, a lip shorter than the sepals, and the midlobe stalked. In spite of these details and others, the variations indicate a difference in habitat conditions rather than a species' distinction. The original plants came from near Joinville in Paraná and flowered in May. We know little else about them at present, including whether the synonymy is accurate.

LOEFGRENIANTHUS

This genus has always been a faded color picture in my mind—the badly printed plate 129 from the old Hoehne *Iconografia de Orchidaceas do Brasil* (1949). This unusual plant was found first by the Danish botanist, Dr. Alberto Löfgren, who was making water color paintings of many Brazilian orchids. He found the plant in the Serra da Mantiqueira in 1896. The specimen was lost or misplaced when it was sent to Cogniaux, the leading authority on Brazilian orchids at the time, and no more was heard of it. A chance meeting of Löfgren with Professor Oakes Ames, who was traveling in Brazil with his wife in 1915, led to the remarkable accidental discovery of a second specimen in the forested slopes of Mt. Itatiaya. The whole story is in Ames's *The Rediscovery of a Lost Orchid,* first published in 1946 and republished as a chapter in his *Orchids in Retrospect* (1948).

Löfgren called the species a new form of *Leptotes* in 1918, naming it after Professor Ames's wife, Blanche. Then Hoehne, the Brazilian orchidologist, realized that it was not a form of *Leptotes.* In 1927 he originated a new genus, naming it after Löfgren, and transferred the species, *blanche-amesiae,* to it. This remains a monotypic taxon, another of those that have the lip base fused to the column to form a small nectary chin. The flowers have an unusual configuration of eight pollinia in two groups—one of six larger, the other of two smaller pollinia—with broad caudicles binding them together.

Blanche Ames's Loefgrenianthus

Brazil

(FIGURE 3-9; PLATE 76)

Loefgrenianthus blanche-amesiae (Löfgren) Hoehne. 1927. *Arq. Bot. Estado São Paulo* 1(4):593, t. 1.

Synonym

Leptotes blanche-amesiae Löfgren. 1918. *Arch. Jard. Bot. Rio de Janeiro* 2:58, t. 22A.

Loefgrenianthus is a plant that hangs as a seldom-branched, elongating, and curving set of small, flexible, rhizomatous stems with leaves only on the last 5–7 apical growths; the base of the plant is just a scramble of bare stem with the leaves fallen. The incremental growths are well-separated along the stems, and the leaves are pendent over the lichens and mosses of the branches or tree trunks on which the plant grows. It apparently forms few adventitious roots, so that those at the base of the original rhizome are important for the anchorage and survival of the individual plant. The small leaves are ovate, elongated, and pointed, the largest measuring 4 × 20 mm. Single white flowers with yellow or reddish markings are produced. The stalk measures about 10 mm, and the flower segments are 8–10 mm.

The lip of the flower, besides forming a little spur from the fusion with the lateral sepals and the column base, has an unusual in-rolled serrate and toothed edge. There is a thickened, keeled-vein callus area on the disc and throat of the lip leading to the nectary. This feature is well shown in the drawings (Fig. 3-9) taken from Hoehne's *Iconografia* (1949). The anther cap and column are dark purple. This is another species crying out for pollinator research (Plate 76).

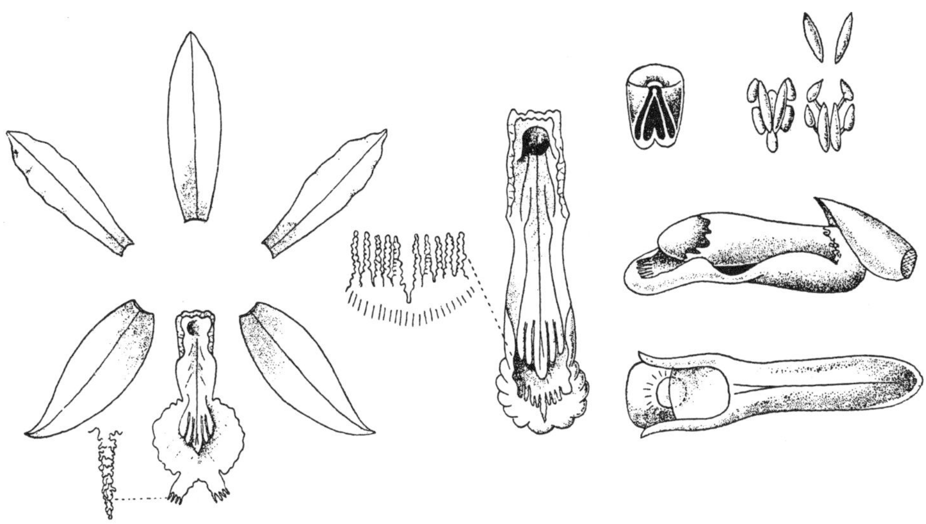

Fig. 3-9. *Loefgrenianthus blanche-amesiae.* Flower detail from Hoehne's *Iconografia* (1949). Permission of the Instituto de Botanica, São Paulo, Brazil.

Brieger and his coauthors in *Schlechter III* tell us that the plant is found at the 1700 m level in the Serra de Mantiqueira and that it is difficult to maintain in culture. It has also been found in Paraná and Rio de Janeiro states, according to Brazilian herbarium specimens checked for me by Antonio Toscano. The plants apparently flower in December or January in Brazil.

There are no previous records of its cultivation in the United States, but I did by chance find a plant in the collection of Phillips and Ann Jesup of Bristol, Connecticut. Apparently, so the story goes, Paul Gripp, who had recently returned to California from a trip to Brazil, took a small piece of his plant to Phil Jesup when he went to speak to the Eastern Orchid Congress in New York. The plant has grown and flowered in the Jesup collection, an assemblage rightly famous for its unusual miniature species of orchids thriving under skillful hands. If other fragments of the original plant survive, they are unknown at present. *Loefgrenianthus* is also apparently unknown in Brazilian nurseries.

NEOLAUCHEA

For some reason *Neolauchea* and its single species is not described in Cogniaux's writings in Martius' *Flora Brasiliensis*. The taxon is described much later in Pabst and Dungs (1975) as a species of *Isabelia*. *Isabelia* and *Neolauchea* have always been so different both vegetatively and florally that one must look for the reasons for combining the two in the same genus. A publication of Senghas and Teuscher, listed below, made the combination. At this point I prefer the old arrangement, so will keep the two genera separate in this book and use the older Kraenzlin name of *Neolauchea* for this monotypic population.

The narrow base of the lip is fused to the base of the column to produce a sac-like nectary, very apparent in a side view of the flower. There are four pollinia. The genus was named by Kraenzlin in 1897 in honor of Director Lauche of the Liechtenstein Botanic Gardens.

Beautiful Neolauchea

Brazil

(PLATE 77)

Neolauchea pulchella Kraenzlin. 1897. *Bull. Herb. Bioss.* 5:110.

Synonyms

Meiracyllium wettsteinii Porsch. 1905. *Oesterr. Bot. Zeitschr.* p. 160.
Isabelia pulchella (Kraenzlin) Senghas and Teuscher. 1968. *Am. Orch. Soc. Bull.* 37:502.

Neolauchea has always been one of my favorite miniature species, with its stiff, branching rhizomes, as often as not projecting straight out into the air, and its many thin, little adventitious roots hanging from them. It looks like a tiny, one-leafed, clambering *Encyclia* until you see the stalks with their single rosy pink or darker flowers. A closer look at the sides of the flowers shows the base of the lateral sepals joining the base of the lip and column, all fusing to form the chin-like enlargement of the nectary. The lip is paler in coloration, and the entrance to the nectary is apparent at its base under the rosy pink column.

The leaves are 5–10 cm tall, and the growths are 2–3 cm apart on the thin rhizomes. The pear-shaped pseudobulbs are less than 1 cm tall, and the flowers are about 10 mm wide. The leaves and pseudobulbs are often flushed with red when the plants are grown in good light.

This plant must be fastened to a slab of cork or hard tree fern. Its exuberant manner could never be retained within a pot. The Hamilton Peak for this species is November. The plants come from Paraná and Rio Grande do Sul.

PINELIA

Pinelia is a genus with tiny plants close to those of *Homolopetalum* in appearance. The latter, however, has proportionately larger flowers. *Pinelia* is centered in South America, and *Homalopetalum* is found in Central America, Cuba, and Jamaica.

Some botanists feel the two genera should be combined, and if that is done *Homalopetalum* will have the priority. I am considering them separately in this series of books. *Homalopetalum,* in turn, is closely related to *Domingoa,* which has a multiple flowering habit over a period of years instead of producing single flowers. These genera are also all related to *Tetramicra* in their details of flower structure.

The genus *Pinelia* is named after the French botanist Chevalier Pinel who found the plants in Brazil during 1853. He thought they looked like miniature cattleyas except for the base of the lip. At Kew is a small water color painting, and the original plant is pasted upon a sheet of paper in the Lindley Herbarium.

The flowers have perianth parts nearly alike in shape, including the lip that dominates the flower and clasps around the column. There is debate over the nature of the lip and its rather large lateral lobings that are curiously attached, apparently to the column base. So few plants have been found, the anatomy of the flowers is not really understood. These lateral lobes may in reality be wings of the column that help make up the walls of the nectary and are not really parts of the lip as such.

The *Pinelia* column is short and fused with the base of the lip to produce the nectary (see Fig. 3-1). The details of anatomy would have to be worked out with a good microscope. There are four pollinia, and there is the possibility of poorly developed extra pollen masses as appendages on them. The genus and species are still poorly known, and I have never heard of any examples of *Pinelia* in cultivation. When the *Homolopetalum* species are eliminated from consideration, there are two species of *Pinelia* in South America and one other in Cuba and Hispaniola. The generic epithet is not to be confused with *Pinellia,* an aroid genus from China and Japan.

Mountain Pinelia

Venezuela

(FIGURE 3-10)

Pinelia alticola Dunsterville and Garay. 1965. *Venezuelan Orchids Illustrated* 3:242–243.

The flowers of *Pinelia alticola* have shorter, broader petals and lip than those of *P. hypolepta*. This tiny plant has globular pseudobulbs to 5 mm height and leaves 5 × 7 mm that are rigid, ovate, and pointed. Single flowers are produced on a stalk 2 cm tall with 5 tiny sheathing bracts and are white with a pink flush. Dorsal sepals are 2.2 mm wide and 3.2 mm long; petals are 1.4 × 3.1 mm; the lip is about 3.8 × 4 mm when flattened from its curved position fused with the column to form a nectary space. There is a transverse callus thickening across the lip near its base that is lacking on drawings of the *P. hypolepta* lip. The plants were discovered by Dunsterville at 1735 m, growing on top of Auyantepui in southeastern Venezuela. Nothing else is known about the species.

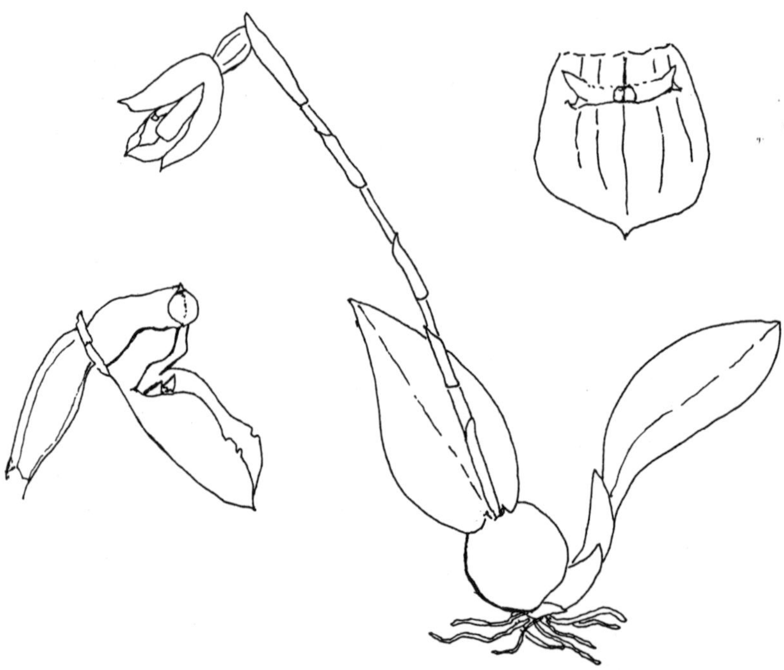

Fig. 3-10. *Pinelia alticola*. After an illustration from *Venezuelan Orchids* (Dunsterville and Garay, 1965). Actual plant size 10–12 mm, flower stalk 20 mm, overall height 30–32 mm.

São Paulo Pinelia

Brazil

(FIGURE 3-11)

Pinelia hypolepta Lindley. 1853. *Fol. Orch.*, part IV.

Synonyms

Restrepia hypolepta (Lindley) Reichenbach f. 1861. *Walp. Ann. Bot.* 6:204.
Pinelia paulensis Hoehne and Schlechter. 1926. *Arq. Bot. Estado São Paulo* 1(3):246–256.

Brieger, Maatsch, and Senghas in *Schlechter III* tell us (p. 508) that these tiny plants come from the coastal mountains of São Paulo; the Alto da Serra is cited by Hoehne. Pinel's plant came from near Rio de Janeiro and is the type specimen for both the genus and the species, *Pinelia hypolepta*.

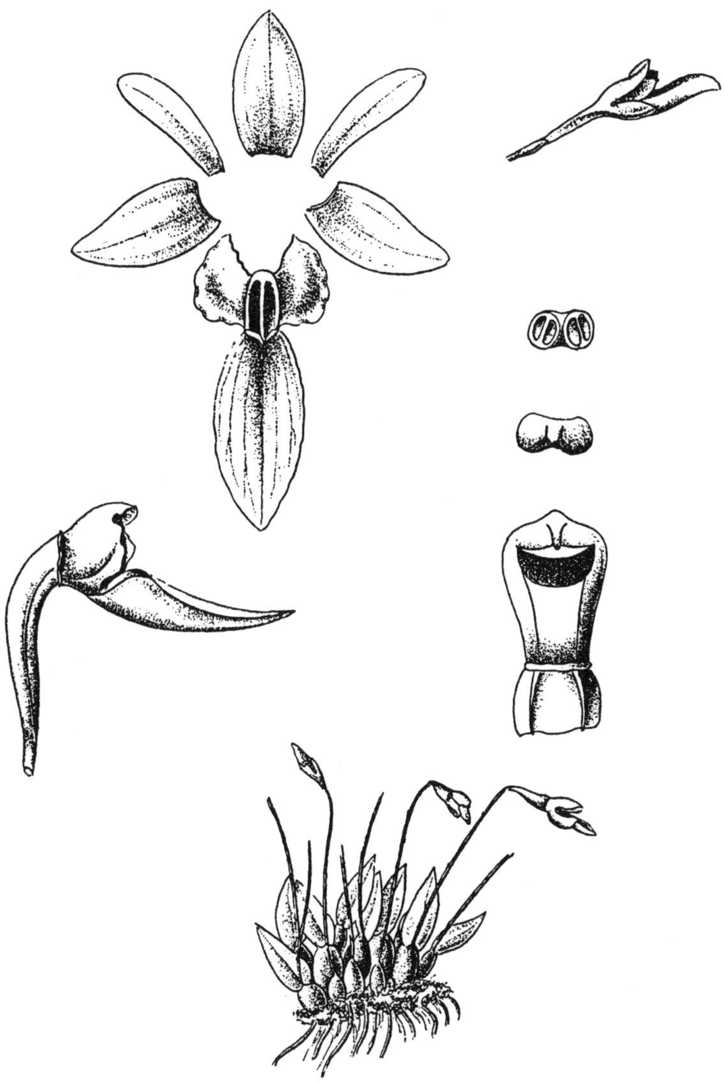

Fig. 3-11. *Pinelia hypoleta*. Illustration from Hoehne's *Iconografia* as *P. paulensis*, plate 59 (1949). Permission of the Instituto de Botanica, São Paulo, Brazil.

Plant size and habit are comparable to that of the plants of *Pinelia alticola,* but the flowers have a different lip structure and callus formation. Reference to the drawing (Fig. 3-11) shows these distinctions to some degree. The plants are only 10–11 mm high, and their flower stalks lack bracts, reach 20 mm and have single flowers.

The tiny flowers are only 6 mm across × 8 mm high and are light yellow with a tinge of green There is a red spot on the anther cap. The column is small, almost horizontal, and has wing-like margins (according to Lindley's type description). Nothing else is known about this species.

Cuban Pinelia

Cuba, Hispaniola

Pinelia leochilus (Reichenbach f.) Garay and Sweet. 1972. *Jour. Arnold Arboretum* 53(3):390–398.

Synonyms

Epidendrum leochilus Reichenbach f. 1865. *Flora,* p. 277.
Pleurothallis tuberculata Wright. 1866. Herb. *ex Griesb. Cat.,* p. 262.
Hormidium leochilus (Reichenbach f.) Cogniaux. 1909. *Symb. Antill.,* p. 472.

Despite the synonomy showing how the botanists were searching for the correct affiliations of this species, little is known about it. Although it is Caribbean, I'm inserting it here with the rest of its genus. It provides another interesting example of Caribbean–South American plant geography to explain and be aware of.

A few details from the type description (all I have available for reference) show it to have a narrowly ovate pseudobulb 5–6 mm tall with one leaf 4–6 × 7–12 mm. The leaf is narrowly ovate and acute. The 2–4 flowers are born on a hair-like stalk 15–25 mm long. The flower parts measure only 1–2 mm and are yellowish green or white. There is a red spot on the anther cap. No illustrations of the species are available, and the color notes come from Wright's observations on the type specimen.

The plants are apparently extremely rare; the recent first-time collection by Alain Liogier in the Dominican Republic near the Mao River made it possible to determine the generic status. It was found originally near Monte Verde in Cuba by W. Wright in 1860. There is no record of plants ever being in cultivation.

DOUBTFUL SPECIES

Lehmann's Pinelia

Colombia

Pinelia lehmanniana Kraenzlin. 1898. *Englr. Jahrb.* 26:479.

This *Pinelia* species has several noticeable bracts along the flower stalk. It is single-flowered and was collected in 1906, growing at 1800 m at Hatico, Colombia. It appears much more like a possible *Homolopetalum pumilio,* but so little is known that we will have to wait for more field work before this species, and the genus, can properly be understood.

PSEUDOLAELIA

Pseudolaelia is a poorly known Brazilian genus, especially outside of Brazil, and the plants have scarcely been grown or flowered in cultivation. Most are high-altitude, cool-growing species difficult to obtain and maintain. They grow usually on scrubby *Vellozia* "trees" in a desert-like mountain habitat. The orchid plants have extended, rambling rhizomes between growths so that the plants scramble over the branches, occasionally producing flowers on an elongated stalk. They do not seem overly floriferous. One gets the general impression of a growth habit like *Schomburgkia* but with smaller-sized plants and flowers. They also appear as though they were related to the Mexican laelias, indicating the logic of their generic name, though the leaves may number 3–7 and are neither rigid nor leathery, as those of their counterparts.

The genus was first established in 1935 by the Brazilian botanists Porto and Brade when they described the species, *Pseudolaelia corcovadensis.* At that time they also transferred Hoehne's *Schomburgkia vellozicola,* named in 1933, to the new genus. Since that time other species have been described—*Pseudolaelia dutrae* by Ruschi in 1949, *P. geraensis* by Pabst in 1967, and *P. cipoensis* and *P. irwiniana* by Pabst in 1973.

Jones (1972a) felt that these were all variants of a hybrid population between *Schomburgkia crispa* and an *Encyclia* such as *E. dichroma.* Today this idea seems to have no basis even though the various *Pseudolaelia* species may not be well separated or their characteristics well understood. Flowers of *Pseudolaelia* have eight equal pollinia. Their center of distribution appears to be in Espírito Santo state. Very little information is available on these species beyond their type descriptions.

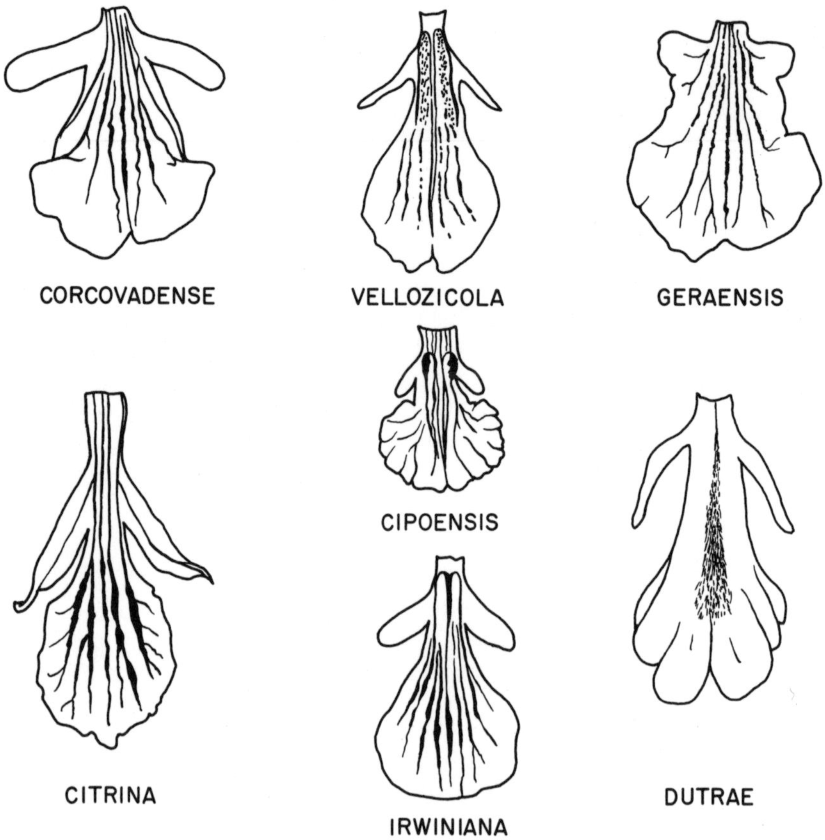

Fig. 3-12. *Pseudolaelia* lip outline drawings. Comparison of the species. After Pabst and Dungs (I, 1975) and others.

Cipó Pseudolaelia

Brazil

(FIGURE 3-12)

Pseudolaelia cipoensis Pabst. 1973. *Bradea* 1(36):365–370, pl. p. 369.

The flower analyses available show *Pseudolaelia cipoensis* very close to *P. corcovadensis,* but there is no real way to compare fresh specimens or give a definitive answer at present. The flowers are violet-purple with a darker purple lip having a white disc. The sepals and petals are broad for their lengths. The lip is fused to the column for about one-third of its length, and the lateral lobes are narrow and point down when the lip is flattened out. Two callus areas are at the base of the lateral lobes. The narrow isthmus expands to a circular, fan-shaped midlobe with a ruffled blade and 3–5 raised veins extending toward the apex.

The plants were collected by Ghillany in 1970 in the Serra do Cipó on *Vellozia*.

Citron Pseudolaelia

Brazil

(FIGURE 3-12)

Pseudolaelia citrina Pabst. 1976. *Bradea* 2(12):65–70.

Synonym

Pseudolaelia luteola Brieger. 1977. Invalidly published name in *Schlechter III.*

This species is yellowish flowered, and it has the smallest plants and flowers in the genus, with sepals to 8 mm long. The racemes have 6–8 flowers. Other flower measurements from the type description are: dorsal sepal, 8 × 16 mm, five-nerved; the lateral sepals are 2 × 8 mm; the lip is 4 mm across the midlobe × 9.5 mm long. The leaves are linear, rigid, and strongly conduplicate, 5 mm wide × 4–7 cm long. Roberto Kautsky discovered the plants in Espírito Santo and also in Minas Gerais near Ibiraçu at 1200 m, where they were growing on *Vellozia.*

Corcovado Pseudolaelia

Brazil

(PLATE 78)

Pseudolaelia corcovadensis Porto and Brade. 1935. *Arq. Inst. Biol. Veg.* 2:207–211.

Pseudolaelia corcovadensis is the type species for the genus and is found growing in Rio de Janeiro state and to the south in the Mantiqueira Mountains. Pabst and Dungs show flowers to have a uniform magenta color with seven raised white veins on the disc. There is a magenta column with a white anther cap. The flowers have broad sepals and petals similar to those of *P. geraensis.*

The plant was first found growing on Corcovado in Rio de Janeiro at 500 m in 1935 by Voll and Cariz. The species is also found in Espírito Santo.

Dutra's Pseudolaelia

Brazil

(PLATE 79)

Pseudolaelia dutrae Ruschi. 1949. *Bol. Mus. Biol. M. Leitão,* Ser. biol., 1:30–55.

The lip of *Pseudolaelia dutrae* flowers has white, narrow, pointed lateral lobes, a creamy yellow center, and a magenta margin with dark veining. There is a dark anther cap. The sepals and petals are rosy pink. This combination of characters seems close to or the same as those of *P. vellozicola,* but there is no way at present to say whether they are the same. I am inclined to make this a synonym under that species but will keep it separate with caveats for the time being. Ruschi states that the difference lies in the narrower, more pointed sepals and petals and a puberulence on the lip base. The size variation in ordinary usage is not in itself enough to count as a species difference, but the puberulence might be. Ruschi's drawings also show a paniculate flower stalk, which can have as many as 120 flowers, while all the other species have simple racemes with only a few flowers. He relates the species to both *P. vellozicola* and *P. corcovadensis*. Nothing else is known.

Pseudolaelia of Minas Gerais

Brazil

(FIGURE 3-12)

Pseudolaelia geraensis Pabst. 1967. *Orquidea* 29:5–13.

The plants of *Pseudolaelia geraensis* produce lilac-colored flowers. The lip is characterized by short, triangular, ear-like lateral lobes. It also has a central sulcus and some raised veins extending to the apex. The petals are broad. There are 5–6 linear-oblong leaves congested at the top of the pseudobulbs, and the inflorescence can reach a height of 90 cm. It appears to be closely related to *P. corcovadensis*. The plants have been found in the Campos region of Minas Gerais, but they are relatively unknown.

Irwin's Pseudolaelia

Brazil

(FIGURE 3-12)

Pseudolaelia irwiniana Pabst. 1973. *Bradea* 1(36):366.

Pseudolaelia irwiniana is closely related to *P. vellozicola,* but the flowers appear to have certain differences, particularly the shorter, thicker column 2.5 mm wide × 4 mm long. The plants were collected by Irwin's group from the New York Botanical Garden on *Vellozia* plants at 900 m in the Serra do Espinhaço near Diamantina in 1972. Nothing more is known about this species.

Vellozian Pseudolaelia

Brazil

(PLATE 80)

Pseudolaelia vellozicola (Hoehne) Porto and Brade. 1935. *Arq. Inst. Biol. Veg.,* 2:207–211.

Synonyms

Schomburgkia vellozicola Hoehne. 1933. *Bol. Agricola de São Paulo* 34: 600–638.

Pseudolaelia auriculata Brieger. 1977. Invalidly published name in *Schlechter III*.

This was the first *Pseudolaelia* species to be described, but Hoehne placed it in the genus *Schomburgkia*. It came from the mountains of southern Bahia. The color plate 938 in Pabst and Dungs shows it as magenta with white throat and disc. It has narrow petals. The column has greenish flushes, and the anther cap appears white with flushed magenta. Its range extends from southern Bahia south through Espírito Santo.

There have been a few plants in cultivation, mostly grown on tree fern slabs or poles. The long rhizomes between growths are covered with stringy bracts.

PYGMAEORCHIS

Pygmaeorchis came to light when the first species was described in 1939. A second species was discovered in 1981. Again, we have almost no information apart from the type descriptions. Generic distinctions include the miniature size of the plants, bifoliate growths, lip fused to the column base, and bicarinate callus. There are eight equal pollinia in two groups of four.

Brazilian Pygmy Orchid

Brazil

(FIGURE 3-13)

Pygmaeorchis brasiliensis Brade. 1939. *Arq. Serv. Forest Rio de Janeiro* 1(1):42.

This miniature is unknown in cultivation and from its color illustration in Pabst and Dungs would seem to be a most unusual curiosity. It has a dark violet-purple corolla that remains semiclosed, and the ovary, covered with papillate protuberances, looks like a tiny gherkin or cucumber. In the Laeliinae this characteristic is found on some *Encyclia* seed pods but is not developed there to the same degree. It is also found on pods of *Constantia microscopica,* another miniature.

The petals are about half the width of the sepals. The lip is hardly differentiated at all, except for two small callus-like enlargements along the veins flanking the central one that runs through the lip. The pseudobulbs are spherical with two oblong-linear leaves. Only a single flower is produced between them, shorter than the leaves.

Plants have been found in Rio de Janeiro in the Organ Mountains at 1500 m near the Pedra do Frade. In Minas Gerais the type collection was made near Maua Itatiaia by Luiz Lanstyak. The species must be closely related to *Constantia,* but there is not really enough herbarium material to tell generic or species limits.

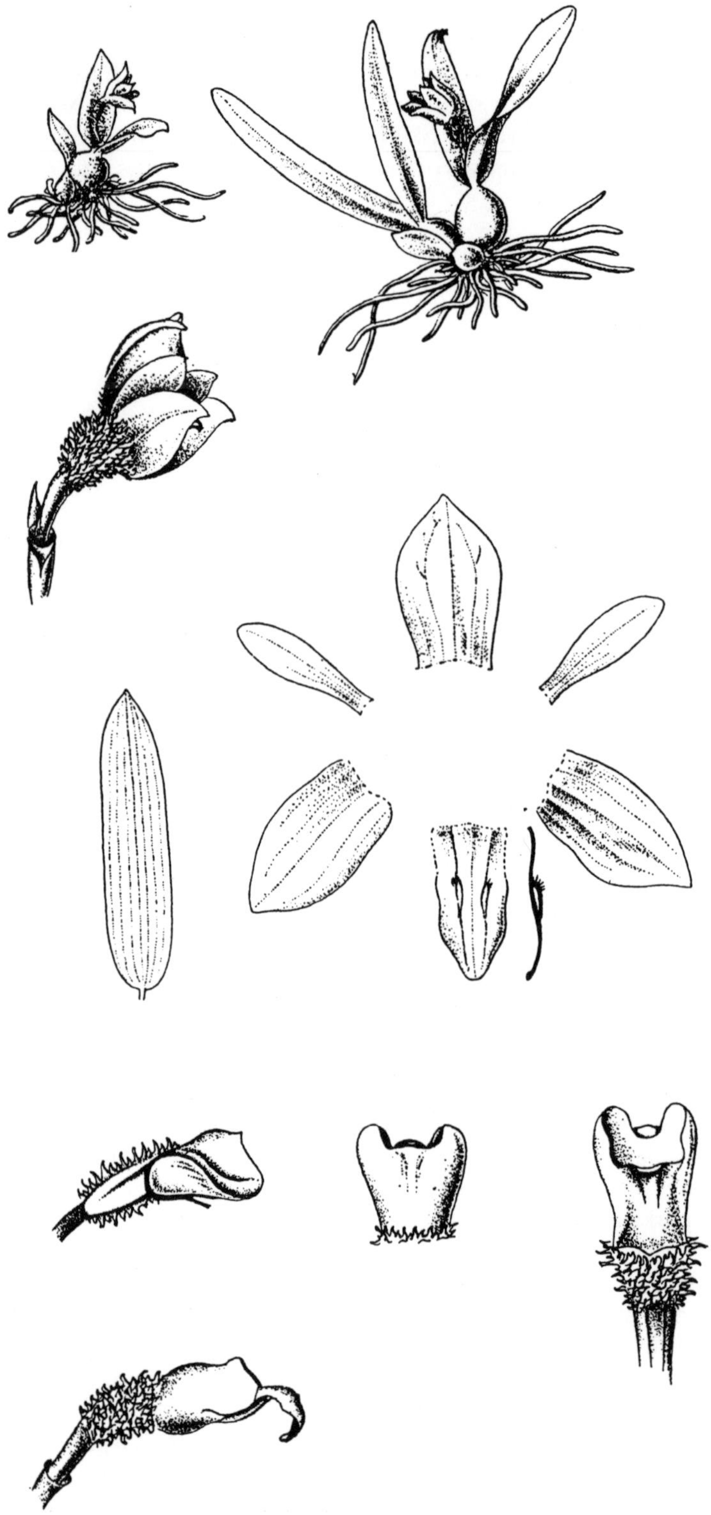

Fig. 3-13. *Pygmaeorchis brasiliensis*. Illustration from Hoehne's *Iconografia,* plate 119-I (1949). Permission of the Instituto de Botanica, São Paulo, Brazil.

Seidel's Pigmy Orchid

Brazil

(FIGURE 3-14; PLATE 81)

Pygmaeorchis seidelii Toscano and Moutinho. 1981. *Bradea* 3(25):193–196.

The main distinction between *Pygmaeorchis seidelii* and *P. brasiliensis* is that the ovary is not covered with papillae in this *P. seidelii*. The plants were collected by Alvim Seidel near Ouro Preto in Minas Gerais. They were growing on *Vellozia*.

The pseudobulbs are 3–7 mm high with the recurved, linear-lanceolate leaves adding 5–15 mm. The erect scape is about 8 mm with a single greenish flower about 4–5 mm long. The ovary is minutely glandular, about 2.5 mm long. The lip is fused to the column one-third of its length, and there are two parallel raised veins on the disc.

Fig. 3-14. *Pygmaeorchis seidelii*. Type illustration from *Bradea* (1981). Permission of the authors and *Bradea*.

REICHENBACHANTHUS

This genus honoring Henri Gustave Reichenbach was described by Barbosa Rodrigues in 1882. In the original account, the word was unfortunately spelled with an *m* instead of an *n* in the conversion to Latin, a misprint perpetuated by Hoehne. The genus was based on the species *Reichenbachanthus modestus,* since found to be a synonym for Reichenbach's 1877 *Hexisea reflexa*. The plant and single leaves are long, mostly pendent, terete, and pointed, the stems hanging with rhizomatous extensions between growths. There are a few scattered papery bracts along the rhizomes, and they also sheathe the bundles of short stalks on which the small, whitish, yellowish or purple stained flowers are produced one at a time. The thickened base of the lip forms a nectary with the base of the lateral sepals, the column, and the column foot, and there are four pollinia.

There is little record of the plants in cultivation, the small flowers being of no interest to date. At least five species have been named, but with synonomy they are reduced to one good species and the possibility of a second or third. These small-flowered, seldom collected, miniature genera all seem to generate more names and confusions than their larger cousins. The genus is related to *Hexisea* and *Nageliella*. Rodrigues placed it between *Ponera* and *Tetragamestus*. Although not mentioned in any of the descriptions, it would appear that these species are repeat bloomers from the same sheath.

Reflexed Reichenbachanthus

Bolivia?, Brazil, Guyana, Suriname

(FIGURES 3-15, 3-16)

Reichenbachanthus reflexus (Reichenbach f.) Brade. 1935. *Rodriguesiana* 1(2):55.

Synonyms

?*Scaphyglottis reflexa* Lindley. 1839. *Bot. Reg.* 25, misc. not. 21.

Hexisea reflexa (Lindley) Reichenbach f. 1877. *Linnaea, Lit. Bericht* 41:131.

Reichembachanthus modestus Barb. Rodrigues. 1882. *Gen. Sp. Orch. Nov.* 2:164.

Fractiunguis brasiliensis Schlechter. 1922. *Anex. Mem. Inst. Butantan* 1(4):56, Pl. 10–11.

Fractiunguis reflexus (Lindley) Schlechter. 1922. *Anex. Mem. Inst. Butantan* 1(4):55–56.

Reichenbachanthus emarginatus Garay. 1967. *Bot. Mus. Leaf. Harvard Univ.* 21:255. (Based on *Hexisia reflexa* Reichenbach f., not on *Scaphyglottis reflexa* Lindley.)

Since there is a confusion over the type specimen of this species, the synonymy of *Reichenbachanthus reflexus* is a real can of worms. Garay believes the Lindley plant (1839) usually cited as the type is in error, is something different altogether, and that the Reichenbach specimen (1877) is the correct type. Both have the same Latin species epithet. When the genus *Reichenbachanthus* was founded by Rodrigues (1882), however, neither of these specimens was cited but, instead, one of his own that he

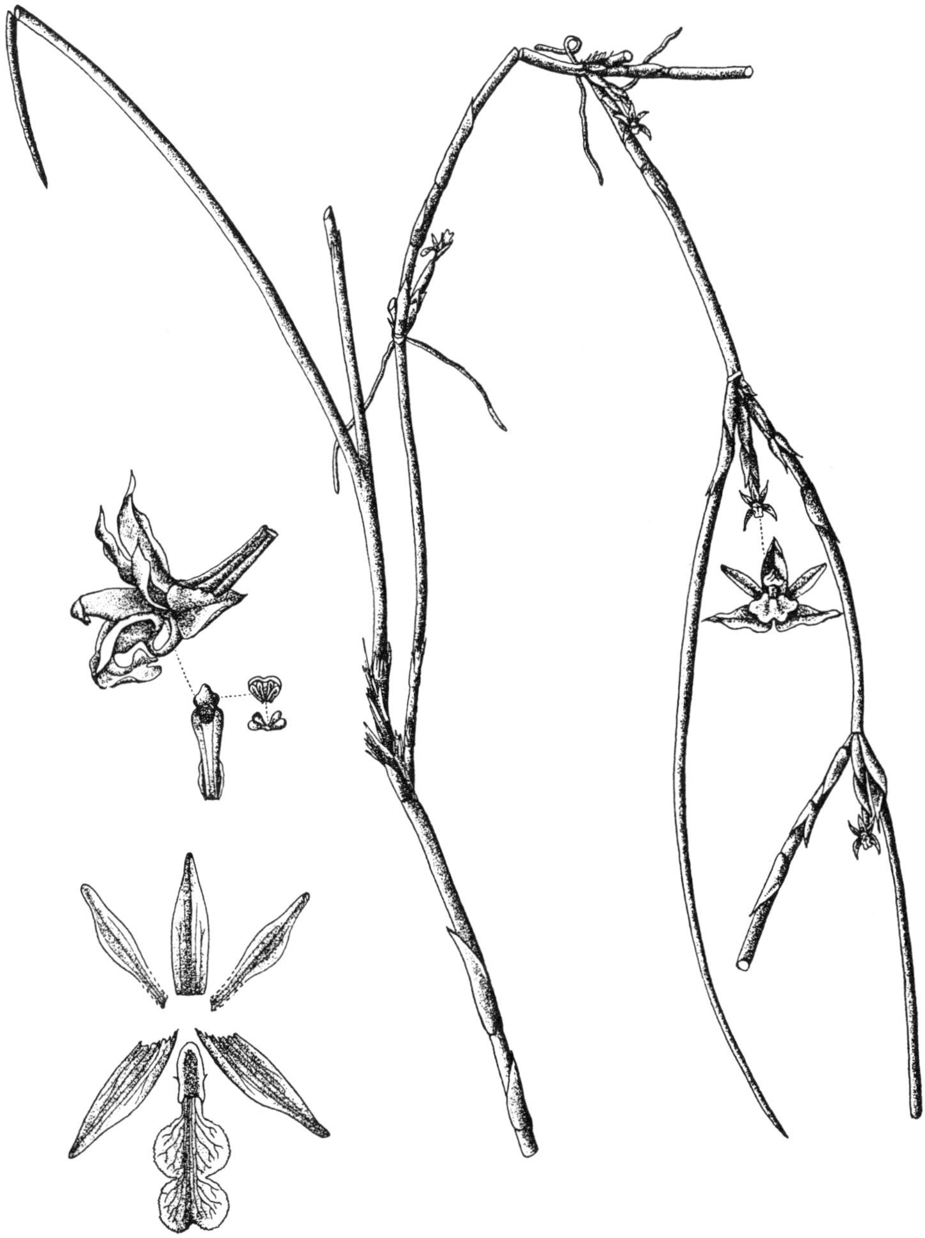

Fig. 3-15. *Reichenbachanthus reflexus* as *R. modestus* (synonym). Illustration from Hoehne's *Iconografia,* plate 101 (1949). Permission of the Instituto de Botanica, São Paulo, Brazil.

called *R. modestus*. Schlechter, in turn, set up the genus *Fractiunguis,* citing only the Reichenbach specimen but then adding one of his own, *F. brasiliensis*. Finally, Garay throws up his hands and says they should all be called by a new name, *Reichenbachanthus emarginatus,* to straighten everything out. Unfortunately, neither of the type specimens of the *Scaphyglottis* or the *Hexisea* that started it all have ever been pictured.

Fig. 3-16. *Reichenbachanthus reflexus*. Plate 290, *Icones Plantarum Tropicarum*, series II, fasc. 3. Permission of the artist and the Missouri Botanical Garden.

Lindley, in the *Botanical Register* under the name *Scaphyglottis reflexa*, describes his specimen:

> A branching, straggling plant, with slender leaves, and solitary, pale, dull-yellow flowers, with a crimson stain along the middle of the lip. Like the rest of the species it is quite destitute of beauty. It was obtained by Messrs. Loddiges from Demerara.

In the Latin description there is mention of semiterete leaves, but nowhere is their pendent habit mentioned. The sepals are described as ovate-triangular and reflexed; the petals are linear and obtuse; and the lip is oblong, emarginate, obtuse, and undulate with the midlobe strongly reflexed. The plant and flower are theoretically well illustrated by Cogniaux in Martius, but whether or not the species pictured is the same thing that Lindley saw is now open to question. Garay says no, and that Reichenbach had a different plant, the correct one.

The Werkhoven book (1986, p. 195) has a good color photo that matches the Cogniaux illustration of *reflexus*, and the plant is what we are calling *Reichenbachanthus reflexus* today. The color of the central veins of the lip appears to be a band of darker, dingy yellow. There are, no doubt, certain variations in coloration among the clones, and that in turn may be related to their geographic distribution. The anther cap is purple, and the stigma is outlined in the same color. The petals are narrower than the sepals.

These plants have 10 cm leaves. The sepals are oval-elongate; the petals are linear-ligulate and narrower than the sepals. The lip is small and violin-shaped, and it extends into an elongated claw that is fused to the column and helps produce the sac-like nectary. The lateral lobes are rounded, and the midlobe is slightly notched. It was collected in Brazil near Rodeio in Rio de Janeiro state, and subsequently in Minas Gerais and São Paulo, flowering in February.

Orchidologists seem to have given this species more than its proper share of names, probably from the lack of research material, the insignificance of the flowers, and the fact that it has been found in so many different places by various collectors. The type specimens from both locations should be reexamined to double check their identities and distributions and ascertain how many taxa there really are—and what their proper basionyms may be.

DOUBTFUL SPECIES

Lankester's Reichenbachanthus

Colombia?, Costa Rica, Venezuela?

Reichenbachanthus cuniculatus (Schlechter) Pabst. 1955. *Rodriguesia* 16/17:138. The same combination was duplicated by Dressler in 1964. *Taxon* 13(7):246–249.

Synonyms

Fractiunguis cuniculatus Schlechter. 1923. *Fedde Repert,* Beihefte. 19:31.
Hexisea cuniculata (Schlechter) Ames. 1925. *Schedulae Orchidianae* 9:48.
Reichenbachanthus lankesteri Ames. 1925. *Schedulae Orchidianae* 9:47.

Ames, in describing *Reichenbachanthus lankesteri* from Costa Rica, says the flowers are white marked with purple at the base of the lip. He also writes that it is closely allied to Schlechter's *Fractiunguis cuniculatus,* also described from Costa Rica, and makes the new combination of *Hexisea cuniculata* from that name. Pabst then makes the combination *Reichenbachanthus cuniculatus* in 1955. This does not help much in deciding whether *R. lankesteri* is something new or just another example of the old *R. cuniculatus,* or even *R. reflexus,* but giving it a very wide, not to mention unusual distribution. Since Schlechter's type specimen no longer exists (it burned in World War II), it is difficult to say exactly whether *R. cuniculataus* and *R. lankesteri* are the same or different. I am inclined to believe they are identical.

The newly published Volume 4 of *Native Colombian Orchids* (Escobar, 1990–1992) casts some light in this direction. There is a fine illustration of a Colombian *Reichenbachanthus* species that could well be *R. cuniculatus* or *R. lankesteri.* The sepals and petals are pale green overlaid with a purplish flush, and the lip is white with three distinct purple markings. One is along the central vein near the apex, and the two others are wider bands extending forward from the base on either side of the central area. The column is green and the anther cap is purple.

The plant is labeled *Reichenbachanthus reflexus* in the Colombian photo, but that identification should be reconsidered. I believe the photo also matches the drawing labeled *Reichenbachanthus reflexus* in *Orchids of Venezuela* by Dunsterville and Garay (1979) and that label should also be reconsidered. In this genus, fresh flowers for study would seem to be more important than the dried herbarium specimens. That is true of other orchid genera as well, especially those with small flowers, a matter that makes species plants in cultivation all the more valuable, particularly when the original country of origin is known.

Reichenbachanthus cuniculataus and *R. lankesteri,* both described initially from Costa Rica, could easily be a Colombian species as well. Plants from Colombian–Costa Rican distribution would not be unusual, and then this taxon could be moved out of the doubtful category.

White Reichenbachanthus

Ecuador

(PLATES 82, 83)

Neither described nor named.

My real problem with this genus is to decide what name applies to the white-flowered forms from Quito, Ecuador. They grow at 500 m and are illustrated in the fine book by Mulder, Mulder-Roelfsema, and Schuiteman (1990), whose slides are reproduced here (see Plates 82, 83). The examples of *Reichenbachanthus reflexus* have been described as yellowish, yellow green, or a dingy color, and it appears so in the Werkhoven illustration. The *R. cuniculatus* forms are white and purple. Do these white forms represent another species or what?

The sepals and petals of the white flowers are more reflexed than those of *Reichenbachanthus reflexus,* and they are also shorter and broader. Both forms have purple anther caps and stigma outlines. More field work and living plants to study will help resolve the complexities within this small genus of one, two, or possibly three species.

RENATA

Ruschi in 1946 set up the monotypic genus *Renata,* which he thought was the result of natural hybridization between *Schomburgkia crispa* and a species of *Epidendrum* such as *E. violascens.* He named the genus after his wife. The plants have up to 14 somewhat soft, strap-like leaves on an elongated pseudobulb, much in the form of a multileaved schomburgkia, and produce a wide-branched panicle with many creamy yellow flowers. New growths before maturity have the appearance of a coarse, rapidly growing reed-stemmed *Epidendrum.* The plants tend to scramble through grassy debris along streambanks or by roadsides and other cleared areas. There are several centimeters of rhizome separating individual growths. The eight pollinia are of equal size. It is now known that this is a separate species, not a natural hybrid. It is amazing that such a large and floriferous species was discovered so late in the orchid history of Brazil, but it has a very limited distribution.

Canaan Valley Renata

Brazil

(FIGURE 3-17; PLATE 84)

Renata canaanensis Ruschi. 1946. *Arq. Publico Estado do Espírito Santo* 3:1–54. Also, 1946, *Orquid. Nov. Estad. Espírito Santo* 5.

The plants of *Renata* grow in a low mountain rain forest on the Atlantic slopes from 300–1000 m. Humidity can be high, and underbrush provides shade cover. These plants are pictured in Ruschi's book (1986, p. 101, 102, 187 and the back cover of the dust jacket), and there is also a colored detail of the flower in Pabst and Dungs (1975, 1: 219, #932). Otherwise, the plant seems to be unknown in the world of cultivated orchids.

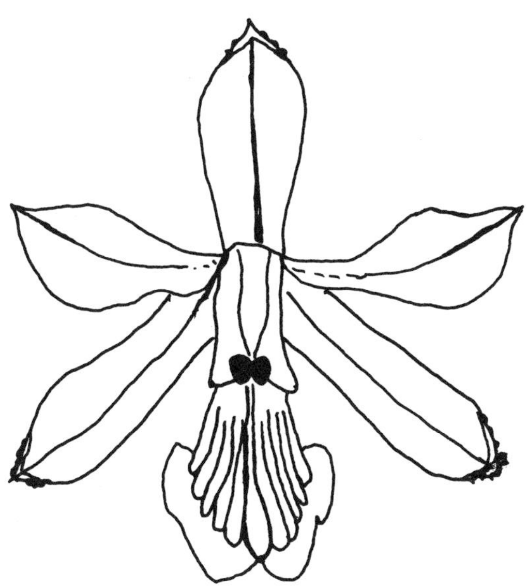

Fig. 3-17. *Renata canaanensis.* Flower detail. After Pabst and Dungs (1975).

The photos of the plants in their native habitat show flower stalks a meter or more in height, with 9–10 wide branches bearing 100 or more flowers. When the flower stalk first appears at the top of the pseudobulbs, it is clothed with a series of loose-fitting bracts. The flowers tend to hang down from the flower stalk branches in a random, non-resupinate fashion. In one herbarium specimen at Harvard the leaves measure 2.2 × 13 cm, the stem is 50 cm, and the flower stalk is 47 cm to the first branch.

The lip has a rather different structure than most Laeliinae and is fused for one-half to two-thirds of the column length. There are much reduced lateral lobes, and the midlobe blade is reduced to a narrow rim around the fore edge of the large, swollen, undulate callus mass. The callus of thickened central veins occupies the isthmus area that makes up half or more of the total lip structure. The midlobe rim is nearly bisected by the indentation at the apex.

This unusual flower cries out for a pollinator study: what insect finds this an attractive landing field and why? Are the flowers really non-resupinate? None of the descriptions mention fragrance. Is it there? We know so little about this species that any new facts would be welcome.

SOPHRONITELLA

Starting out as a *Sophronitis,* this species was transferred to its own genus in 1920 by Schlechter. It never fit well as a *Sophronitis;* it has a different column and flower structure, not to mention the two dark green, narrow, channeled, linear leaves atop the similarly dark, longitudinally ridged pseudobulbs. As orchidists learned more about *Sophronitis* species, the separation of *Sophronitella* and *Constantia* from them was only logical. The flowers have eight pollinia.

Violet Sophronitella

Brazil

(PLATE 85)

Sophronitella violacea (Lindley) Schlechter. 1925. *Fedde Repert,* Beihefte 35:76.

Synonyms

Sophronitis violacea Lindley. 1840. *Bot. Reg.* 26, misc. 18.
Cattleya violacea (Lindley) Beer. 1854. *Prakt. Stud. Fam. Orch.,* p. 209 (*non*-Rolfe).

These plants require more light than *Sophronitis* and tolerate more dryness. Some of the best plants I have seen have been mounted on cork—a satisfactory arrangement if there is sufficient humidity in the growing area. Figure 250 in *Native Orchids of Brazil* (1977) shows the plants in nature, growing on the lichen-covered, corky, and deeply furrowed bark of some oak-like tree in a bright, sunny, and not-too-wet location. In nature they also grow on rock, as was described in Leinig's account (1971) of *Isabelia*.

The clone 'Fox Den' received a cultural award from the American Orchid Society when it had 27 flowers and 5 buds on 15 inflorescences. The natural flower spread was 3.2 cm horizontally and 3.0 cm vertically. The two or three flowers per stalk in cultivation are more than the number usually seen on wild plants, which tend to produce only single flowers. This may be one of those unusual orchids that does better in cultivation than in the wild. It is adapted for survival in its habitat, not for maximum flower production. The cultivar 'Jewels' had 37 flowers and 7 buds on 12 inflorescences, and the natural spread of the flowers was 4.3 cm—obviously deserving the cultural certificate and the AM flower award it received from the American Orchid Society.

The flower color does not seem to vary at all in the clones seen at shows and in collections. It is a glowing fuchsia-violet-purple color on the corolla, and the column is a deeper purple color. White (or *alba*) clones have been described, originally by Barbosa Rodrigues in 1882, but to my knowledge these are not displayed today. In Brazil the flowering is through July and August, but Hamilton's Peak falls in December and January for plants in cultivation.

This species, in a rocky habitat called Vila Velha in Paraná, has formed a natural hybrid with *Isabelia virginalis*, which also grows there on the sandstone. The hybrid, *Isanitella pabstii*, was named by Professor Milton Leinig of Curatiba and is discussed above in the account of *Isabelia*.

Selected References for Additional Information

Adams, C. Dennis. 1971. Orquideas de las Islas Cayman. *Orquideológia* (Colombia) 6(3):144–147.

American Orchid Society. 1970–1989. *Awards Quarterly*, Vols. 1–20.

Ames, Oakes. 1948. *Orchids in Retrospect*. Botanical Museum of Harvard University. Cambridge, Massachusetts.

Ames, Oakes and D. Correll. 1953. *Orchids of Guatemala*. Fieldiana: Botany 26:1, 2. Field Museum of Natural History. Chicago, Illinois.

Assoc. Orquidófila de São Paulo. 1977. *Native Orchids of Brasil*. Miura Mericlone Ltd. Noborito, Kawasaki, Japan.

Bentham, G. and J. D. Hooker. 1883. *Genera Plantarum* 3:526–535.

Brieger, F., R. Maatsch, and K. Senghas. 1976–1981. Third edition of R. Schlechter's *Die Orchideen*, Lieferung 8–12 (*Schlechter III* in text). Verlag Paul Parey. Berlin and Hamburg.

Carnevali, G. 1982. Una nueva especie de *Schomburgkia* para Venezuela y la ciencia. *Ernstia* 10:1–9.

Cogniaux, A. 1898–1902. Orchidaceae II, Tribus VII Laeliinae. In Martius' *Flora Brasiliensis*, Vol. 3, part 5 (Martius in text).

Cogniaux, A. 1906. *Bull. Soc. Roy. Bot. Belge*. 43:325.

Cogniaux, A. and A. Goossens. 1896. *Orchidées. Dictionnaire Iconographique*. Brussels, Belgium.

Dodson, C. H. 1988. Una lista de especies de orquídeas reportadas del Ecuador. In *Orquídeas de la Costa del Ecuador*, A. Arosemena et al. Guayaquil, Ecuador.

Dodson, C. and P. Dodson. 1980. *Icones Plantarum Tropicarum. Orchids of Ecuador*. The Marie Selby Botanical Gardens. Sarasota, Florida.

Dodson C. and R. Vásquez Ch. 1989. *Icones Plantarum Tropicarum. Orchids of Bolivia*. Series II. 2(4):302. Missouri Botanical Garden.

Dressler, R. 1967. The genera *Amblostoma, Lanium,* and *Stenoglossum* (Orchidaceae). *Brittonia* 19:237–243.

Dressler, R. 1981. *The Orchids, Natural History and Classsification.* Harvard University Press. Cambridge, Massachusetts.

Dunsterville, G. and L. Garay. 1959–1976. *Venezuelan Orchids Illustrated,* 6 vols. Andre Deutsch, Ltd. London.

Dunsterville, G. and L. Garay. 1979. *Orchids of Venezuela: An Illustrated Field Guide,* 3 vols. Botanical Museum of Harvard University.

Duveen, Denis I. 1975. Some novelties among Brazilian orchids. Part V. *Laelia esalqueana. Orchid Digest* 39(2):67–69.

Duveen, Denis I. 1977. A new found Organ Mountain habitat of *Sophronitis mantiqueirae. Orchid Digest* 41(3): 104–105.

Engler, A. and K. Prantl. 1889. *Die Natürlichen Pflanzenfamilien.* Part II. Engelmann. Leipzig.

Escobar, Rodrigo, coordinator. 1990–1992. *Native Colombian Orchids.* 4 Vols. Colombian Orchid Society. Medellín, Colombia.

Foldats, E. 1970. *Flora de Venezuela.* Vol. 15, part 3. Orchidaceae. Instituto Botanica. Caracas, Venezuela.

Fowlie, Jack A. 1963. Ecology notes: *Schomburgkia lyonsii. Am. Orch. Soc. Bull.* 32(1):31–33.

Fowlie, Jack A. 1968. *Sophronitis coccinea. Orchid Digest* 32(9):272–273.

Fowlie, Jack A. 1972. A contribution to a further clarification of the genus *Sophronitis,* including the elevation to full specific status of two previously described varieties. *Orchid Digest* 36(5):181–194.

Fowlie, Jack A. 1975a. Rediscovering *Laelia esalqueana. Orchid Digest* 39(2):70.

Fowlie, Jack A. 1975b. With Ghillany in Brazil. Part VIII. Without Ghillany—refinding the "lost" *Oncidium marshallianum. Orchid Digest* 39(2):75–78.

Fowlie, Jack A. 1975c. With Ghillany in Brazil. Part IX. A new *Sophronitis* from Pico de Açu, *Sophronitis acuensis. Orchid Digest* 39(3):147–151.

Fowlie, Jack A. 1977. Some notes on the vegetative appearance of the various *Sophronitis* species. *Orchid Digest* 41(3):106–108.

Garay, L. and C. Slocum. 1963. A note on *Schomburgkia Brysiana. Am. Orch. Soc. Bull.* 32(5):373–374.

Greuter, W. et al. 1988. *International Code of Botanical Nomenclature.* Koeltz Scientific Books. Königstein, Germany.

Haffer, Jürgen. 1969. Speciation in Amazonian forest birds. *Science* 165(3889):131–137.

Hamer, Fritz. 1974, 1981. *Las Orquídeas de El Salvador,* 3 vols. Ministerio de Educacion. San Salvador, El Salvador.

Hamer, Fritz. 1982. *Icones Plantarum Tropicarum. Orchids of Nicaragua.* The Marie Selby Botanical Gardens. Sarasota, Florida.

Hamilton, Robert M. 1986. *When Does It Flower?* Second edition. Published by the author. Vancouver, B.C.

Hoehne, F. C. 1949. *Iconografia de Orchidáceas do Brasil.* Secretaria de Agricultura. São Paulo.

Hoehne, F. C. 1952. Algo concernente ao genero *Laelia* Lindl. das orchidáceas e um nova especie para ele do estado de Minas Gerais. *Arquivos de Bot. do Estado de São Paulo* 2(6):157–167.

Horich, C. K. 1988. Anmerkungen über *Schomburgkia tibicinis* Bateman und *Schomburgkia brysiana* Lemaire, zwei mittleamerikanische "Ameisen-orchideen." *Die Orchidee* 39(5):169–173.

Jones, H. G. 1963. Studies in *Schomburgkia*. *Am. Orchid Soc. Bull.* 32(1):4–12.

Jones, H. G. 1965. *Schomburgkia tibicinis* Bateman. (Orchidaceae) and its varieties. *Adansonia* 5(1): 41–48.

Jones, H. G. 1966. Some notes on the genus *Schomburgkia* Lindl. of the Orchidaceae and its relationships. *Ann. Naturhist. Mus. Wien* 69:57–67.

Jones, H. G. 1967a. Nomenclatural notes and new concepts in the genus *Schomburgkia* Lindl. of the Orchidaceae. *Fedde Repert.* 76(3):193–199.

Jones, H. G. 1967b. A new *Schomburgkia* from the West Indies, with notes on the related species. *Webbia* 22 (1):225–231.

Jones, H. G. 1967c. Eine Bemerkung zu *Schomburgkia brysiana* Lemaire. und ihrer Varietaten. *Die Orchidee* 18(6):298–300.

Jones, H. G. 1968a. The genus *Schomburgkia* in México and Central America. *Ceiba* 14:27–37.

Jones, H. G. 1968b. Studies in neotropical orchidology. *Acta Bot. Acad. Scient. Hungaricae Tomus* 14(1–2): 63–70.

Jones, H. G. 1968c. Notes on a collection of orchids from the West Indian island of Trinidad. *Candollea* 23(2): 295–299.

Jones, H. G. 1968d. Synopsis of sections and specific groups in the genus *Schomburgkia* (Orchidaceae). *Israel Jour. Bot.* 17:196–199.

Jones, H. G. 1969. Nomenclatural notes on some Peruvian orchids. *Publ. del Mus. de Hist. Nat. "Javier Prado"*. Bot. Ser. B 24:1–5.

Jones, H. G. 1970a. A note on the *Schomburgkia undulata* complex. *Orchid Review* 78:110–111, 115.

Jones, H. G. 1970b. Orchidaceae americanae. *Caldasia* 10 (50):491–495.

Jones, H. G. 1972a. Sinopsis del genero *Schomburgkia* en Colombia. *Orquideológia* (Colombia) 7(1):7–12.

Jones, H. G. 1972b. A note on the nomenclature and typification of the *Schomburgkia* species of the Greater Antilles. *Norw. Jour. Bot* 19(2):107–109.

Jones, H. G. 1972c. Notes on some natural hybrid orchids of tropical America and the West Indies. *Bol. da Soc. de Broteriana* 46:239–252.

Jones, H. G. 1973. The genus *Schomburgkia*, a study in the history and bibliography of plant taxonomy. *Taxon* 22(2–3):229–239.

Jones, H. G. 1974. Notes on South American orchids. *Lilloa* 34(1):5–13.

Jones, H. G. 1976. Nomenclatural revision of the West Indian species of *Schomburgkia* (Orchidaceae). *Bull. Torrey Bot. Club* 103(3):117–121.

Kennedy, George. 1975. *Laelia autumnalis* Lindley and *Laelia gouldiana* Reichenbach. *Orchid Digest* 39 (2):71–73.

Kennedy, George. 1979. The genera *Schomburgkia* and *Myrmecophila*. *Orchid Digest* 43(6):204–211.

Kenny, Julian. 1988. *Orchids of the Eastern Caribbean.* Macmillan Publishers Ltd. London.

Kraenzlin, F. 1911. Beiträge zur Orchideenflora Südamerikas. *Kungl. Svenska Vetenskaps Handl.* 46(10):55, t. 3.

Leinig, Milton. 1971. *Isanitella*—a new intergeneric hybrid. *Am. Orch. Soc. Bull.* 40:709–712.

Lindley, J. 1830–1840. *The Genera and Species of Orchidaceous Plants*. Ridgways. London.

Lindley, J. 1837–1842. *Sertum Orchidaceum, A Wreath of the Most Beautiful Orchidaceous Flowers*. James Ridgway and Sons. Picadilly, London.

Martius. See Cogniaux 1989–1902.

Mee, Margaret. 1988. *In Search of Flowers of the Amazon Forest*. Nonesuch Expeditions, Ltd. Woodbridge, Suffolk, England.

Mulder, D., T. Mulder-Roelfsema, and A. Schuiteman. 1990. *Orchids Travel by Air*. Het Houten Hert. Wageningen, Netherlands.

O'Brien, James. 1886. *Sophronitis violacea. Gard. Chron.*, n.s., 18:492.

Pabst, G. 1977. *Schomburgkia gloriosa. Bradea* 2(24):168.

Pabst, G. and F. Dungs. 1975, 1977. *Orchidaceae Brasilienses*, Vols. I and II. Brücke-Verlag Kurt Schmersow. Hildesheim, Germany.

Pessoa, Alvaro. 1983. *Sophronitis coccinea* in its natural habitat. *Orchid Digest* 47(2):45–50.

Pessoa, Alvaro. 1988. Em busca de duas especies perdidas de sophronitis brasileiras. *Boletim CAOB* 1:8–11.

Reichenbach, H. f. 1861. The genus *Bletia*. *Walpers Ann. Bot. System.* 6:418–445.

Reinikka, Merle A. 1972. *A History of the Orchid*. University of Miami Press. Coral Gables, Florida.

Rodrigues, J. Barbosa. 1868–1873. *Iconographie des Orchidees du Bresil*. Six volumes of water color paintings.

Rodriguez Caballero, R. L. 1986. *Generos de Orquídeas de Costa Rica*. Editorial de la Universidad de Costa Rica. San José, Costa Rica.

Rolfe, R. 1913. *Schomburgkii wallisii*. *Orchid Review.* 21:154.

Rolfe, R. 1917. The genus *Myrmecophila*. *Orchid Review* 25:49.

Ruschi, Augusto. 1986. *Orquídeas do Estado do Espírito Santo*. Expressão E Cultura. Rio de Janeiro, Brazil.

Sanders. *Sanders' Orchid Guide.* Revised 1927 ed. St. Albans, England.

Schlechter, R. 1913. Die Gattung *Schomburgkia* Lindley. *Orchis* 7:38–43.

Schlechter, R. 1917. Die Einteilung der Gattung *Laelia* und die geographische Verbreitung ihrer Gruppen. *Orchis* 11(5):87–96.

Schlechter III. See Brieger et al. 1976–1981.

Schweinfurth, Charles. 1959. *Orchids of Peru*, Vols. 2, 3. Fieldiana 30(3), 1960. Field Museum of Natural History. Chicago, Illinois.

Schweinfurth, Charles. 1967. Orchidaceae of the Guayana Highland. *Mem. N.Y. Bot. Gard.* 14(3):69–214.

Sheehan, Tom and M. Sheehan. 1968. Orchid genera illustrated—VII. *Am. Orch. Soc. Bull.* 37(7):581.

Smith, Lyman B. 1962. A synopsis of the American Velloziaceae. *Contrib. U.S. Natl. Herb.* 35(4): 251–292, plates 1–12.

Spencer, Ralph. 1968. "Suffering" sophronitis. *Orchid Digest* 32(9):268–271.

Teuscher, H. 1968. *Isabelia virginalis* and *I. pulchella*. *Am. Orch. Soc. Bull.* 37(6):502–503.

Van der Pijl, L. and C. Dodson. 1966. *Orchid Flowers, Their Pollination and Evolution*. University of Miami Press. Miami, Florida.

Vásquez Ch., R. and C. Dodson. 1982. *Icones Plantarum Tropicarum*. *Orchids of Bolivia*. The Marie Selby Botanical Gardens. Sarasota, Florida.

Veitch, J. and Sons. 1887. *Manual of Orchidaceous Plants, Cattleya*. Chelsea, London.
Werkhoven, Marga C. M. 1986. *Orchids of Suriname*. VACO N. V. Uitgeversmaatschappij. Paramaribo, Suriname.
Wiard, Leon. 1987. *An Introduction to the Orchids of México*. Cornell University Press (Comstock). Ithaca, New York.
Williams, Louis O. 1941. The validity of the genus *Schomburgkia*. *Darwiniana* 5:74–77.
Williams, Louis O. 1946. Flora of Panama III, fasc. 2–5. *Ann. Mo. Bot. Gard.* 33(1)–36(2):107–593.

Plant Name Index

Index of Persons

General Index